WATER SUPPLY SYSTEMS AND EVALUATION METHODS

Water Supply Systems and Evaluation Methods

Rose Marie O. Mendoza

ARCLER PRESS

www.arclerpress.com

Water Supply Systems and Evaluation Methods

Rose Marie O. Mendoza

Arcler Press
2010 Winston Park Drive,
2nd Floor
Oakville, ON L6H 5R7
Canada
www.arclerpress.com
Tel: 001-289-291-7705

001-905-616-2116

Fax: 001-289-291-7601

Email: orders@arclereducation.com

ISBN: 978-1-77361-225-6 (Hardcover)

About the Author

Rose Marie O. Mendoza

Dr. Rose obtained her PhD in Chemical Engineering from the University of the Philippines-Diliman in 2013. She is also a Professor in the Graduate School Department under the Master of Engineering Program at Adamson University since 2006, and is a Visiting Research Fellow at the Department of Environmental Engineering and Science, and the Department of Environmental Resource Management in Chia Nan University of Pharmacy and Science, Tainan Taiwan since 2010. She is now taking up her Post Doctorate Degree in Green Power: Hydrogen Generation and Fuel Cell Development at the University of Philippines-Diliman under the Department of Mining, Metallurgy and Materials Engineering and the University of California, Merced, USA under the Department of Mechanical Engineering.

Contents

List of Figures

List of Tables

Preface

Water supply is one of the most basic requirements of life. As inscribed in the WHO (2006) "Water for Life" Act, it is each and everybody's basic right to have access to safe and efficient water supply. Though the act focuses on drinking water, it is implied that the access to safe and efficient water supply must be made available to all, particularly by the government and managing agencies. Water helps in protecting the health of people, improving the quality of life, and stimulating development of economy, both in rural and urban areas. It is believed that adequate water supply is one of the internationally recognized indicators of economic development, public health, and country's competitiveness. However, the majority of people around the world are living without safe and clean water supply. Studies have shown that over 80% of the total cost incurred in water supply system is tailored towards transmission and distribution of water. It is of greatest benefit therefore that the principles of effectiveness, efficiency, affordability, sustainability, and appropriateness should be adhered to, to improve water supply systems both in rural and urban areas.

Today, several developing countries are challenged with water supply scarcity. Water supply challenges in urban are mainly attributed to the rapidly growing population and increased urbanization. Researches show that the need for clean and safe water has put great pressure on various public institutions because access to this precious commodity is usually faced with a lot of challenges. Some of the major challenges to accessing clean and safe water include risks of contamination from sewage discharge, farm chemicals, bacteria and harmful parasites.

This book has been prepared in order to equip the reader with the basic understanding of the evaluation of water supply system management. It has integrated different concepts related with applicable water supply management. It also presents a number of basic principles applied in water supply systems.

This book caters for the needs of all graduate and undergraduate students taking courses in water supply systems management, various researchers, governmental professionals involved with water supply systems related functions, and various practising engineers. The author has used the simplest and understandable language in order to assist the students, researchers, engineers, and water supply managers to properly conceptualize the material presented. Although there are some references presented at the end of the book to offer essential information to the optimization evaluation of water supply system management, it is worth noting that this book has been prepared using these references in order to provide a comprehensive and detail information.

There are ten chapters in this book and each presents a specific concept with regards to water supply system management. Chapter one discusses the introduction to water supply systems and evaluation methods. This chapter mainly focuses on basic information on water supply systems information and various concepts used in water supplying activities. Chapter two talks about details on evaluation of piped supply system and ground water quality. Chapter three mainly discusses on various ways of improving water security in arid and semi-arid areas through rainwater harvesting. One of the important concepts that are discussed entirely in this chapter is various types of rainwater harvesting and storage structures. Chapter four touches the issues on management and governance of water supply systems in the informal settlements. One of the areas deeply considered in this chapter are the factors influencing water supply in informal settlements. Chapter five discusses about financial evaluation on small-scale water supply investments. Chapter six tackles the basics on evaluation of women's role in water supply systems which stresses on the various problems that women face in supplying water in their household and ways that can be put into practice to reduce these problems. Chapter seven presents a detailed discussion on privatization of water supply services. Chapter eight, chapter nine, and chapter ten encompasses management of non-revenue water (unaccounted for water); threat to potable drinking water; and water crisis, respectively.

Author

1 INTRODUCTION TO WATER SUPPLY SYSTEMS AND EVALUATION METHODS

INTRODUCTION

For a long time water has been perceived as one of the significant natural resources because of its value in promoting prosperity and wealth in the world. For the last few decades, different countries around the world have witnessed an emerging issue of falling water tables and aquifers getting depleted. Additionally, water systems and resources are failing to sustain the rapidly growing population in the world due to the threats posed by pollution arising from various human activities.

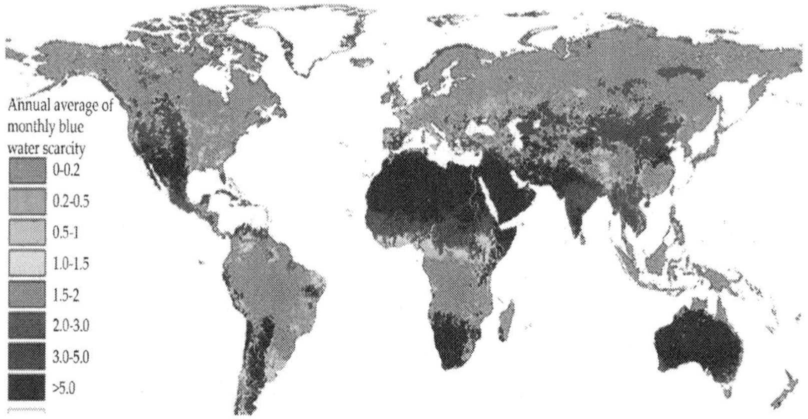

Source: http://www.huffingtonpost.com/entry/water-scarcity-study_us_56c1ebc5e4b0b40245c72f5e

Figure 1.1: Distribution of the global population not served with improved water supply, by region.

According to the American Water Association (1999), about 15% of the overall world's population (approximately to 800million people) still lack access to clean and safe water. The world population will continue to grow and will not only affect the demand for water but also its cost. For this reason therefore, per capita water use is expected to decline with higher population densities. It is well documented that decrease in access to clean, safe, adequate, and quality water is likely to increase the rate of poverty in the society.

Studies have also shown that less than 15 states worldwide have about 50% access to clean and safe water, and this suggests that even though water is abundant, its availability and accessibility for drinking, agriculture, and also for industrial use are limited and unevenly distributes geographically. However, very little emphasis has been paid to modern scientific approaches that protects various water resources and will replenish those that have been depleted.

Adequate water supply enormously contributes to economic productivity and social well-being of the human population. According to a study conducted by UNEP in 2000, about 25% of people in the world lack access to clean and safe drinking water, another 55% of people lack access to safe sanitation systems hence leading to serious water shortages and waterborne related diseases. Therefore, with the rapidly growing world's population (which is estimated at 80 million of people per year), there is the need to add about 65 billion M^3 of water annually.

Urbanization and industrialization and the water-intensive lifestyles have greatly contributed to global water crises. This has been attributed to the rapidly growing population which causes changes in the socio-economic activities of the people. In spite of the crucial nature of access to quality water demand and distribution for sustainable development and improved livelihoods, its decline is becoming a worldwide problem. For instance, studies have indicated that only one third of the population in African countries have access to improved water facilities while the rest are water stressed.

High water demands and lack of effective and sustainable strategies for addressing natural and man-made problems affecting water resources including climate variability projects Africa to be a physically and economically water stressed by 2025. However, current reports indicate that about 58% of the people living within the world's major basins have no access to portable water, making them vulnerable to extreme water shortage despite the targets set by different governments to have water for all.

In a country like Kenya, water demand and availability in major basins is reported to have decreased due to a high average population growth rate of 7%, increased demand for food, rapid economic growth and climate change. This has led to many other challenges such as inability to meet the basic human need of 25 litres of water per capita per day for all people. In addition, the low flows of most rivers in the world are on average maintained below 80% at all times, and this represents a threat to ecologically sustainable development and livelihoods, absolute water scarcity and subsequent conflicts.

Many communities living within major river basins experience water problems rooted in deteriorating access to quality water, unreliable availability of water and the inefficiency of water providers. Women and young girls bear the burden of water collection usually in unsafe environments, making them vulnerable to sexual assault and accidents. The time and energy spent walking and hauling water from distant sources and the resulting water borne diseases keeps them from school work and forcing them to get confined with care of their families. Water is central to the conservation of ecosystems and also for the development of health, agriculture, power generation, livestock production and other important economic activities.

Source: https://www.fluidtime.com/de/aktuelles/Weihnachts-W%C3%BCnsche-2014

Figure 1.2: Women bear the burden of water collection usually in unsafe environments

The increasing competition for domestic water use worsened by inconsistent water supply and the ever- mounting demands from other water sectors has led to an abuse of most water resources worldwide. For example, people use rivers unhygienically (e.g. to bathe and wash clothes) while point pollution (for instance, from cattle dips along river basins) greatly affect water quality. These water challenges call for the need for new water demand management strategies and expansion of water distribution networks.

The inadequate access to quality water in most places is a priority problem just as water resource pollution. Inadequate knowledge in rain water harvesting techniques has played an important role in contributing to water shortage in most places around the world. In some places, especially in developing countries, water is fetched mainly using donkeys usually by women and children. This is a threat that has pushed many communities into extreme poverty. Efficient management of water resources to benefit rural and urban communities requires a full understanding of existing patterns of water demand.

Demand analysis is an important tool for the economic analysis of household behaviour with regard to water use. The rapid growth of urban and rural population together with the increasing growth of mega towns and cities with expansive and intensified agricultural and industrial activities has resulted in economic water scarcity which deters access to meet human demands.

A study conducted by the United Nations Development Programme (UNDP) in the year 2010 revealed that decision-making in water resources is usually dominated by knowledgeable individuals and this excludes the vulnerable groups such as women and the poor in the community, who often lose out in the process of water resource formulation because they lack the resources (knowledge, time, travel and money required to obtain formal authorisation). Little guidance is given in the legislation on how equitable distribution of water is applied to households and other sector demands. Furthermore, inequitable allocation of water is contributed by poor response to water rights, lack of transparency and accountability, especially in areas poorly financed by the government together with corrupt officials and poor rehabilitation of water catchments. The fact that local regulations for Environmental Flows is lacking, monitoring and controlling of water resources has not been thoroughly implemented in most countries worldwide. The different activities such as deforestation, water abstraction for livestock and agricultural irrigation among others are heavily affecting the ecosystem, but the major cause of deteriorating access of quality household water demand from improved sources remains unknown.

THE SITUATION OF WATER SUPPLY IN DEVELOPING COUNTRIES

The challenge of community development in many communities specially in developing countries has been hastened by inadequate supply of safe water. Though the problem of water supply have been neglected for long by the assumption that there is available water in most places, the fact is that, only a small portion of people have access to clean water. Many people especially women and children are therefore forced to walk for long distances in the search of safe water. This exercise is not only tiresome but also burdensome. This has caused people to use water from contaminated sources endangering human health.

Though the world is dominated by almost 75% of water, the hydrogeological distribution of water all over the world is not even. Some parts of the world are well watered by natural rivers flowing from different water catchment area while some parts remain dry almost all through the year.

Water service providers have been facing various challenges hindering them from supplying adequate and efficient water to people in developing countries. This has led to the stalling of established water projects or failure to implement otherwise designed projects.

Most of the people depend on rain water for domestic water and for watering their farms. Many times, the rains have been unpredictable making many people run at a loss again. The available sources of water in most places, especially in arid and semi-arid places are far from the households making the women and children travel far distances to fetch water. This utilizes a lot of time which could have been otherwise engaged in other and more important economic activity for further development of the community and the country. The rainfall patterns are usually unpredictable and hence cannot be relied upon for farming and use. Most of the people in developing countries depend highly on agricultural activities and a reliable supply of safe water is very essential for irrigation.

A reliable and efficient water service will ensure that the rich agricultural zone continues to feed the nation. It is therefore much necessary to study the water services supply situation, the efficiency of the water service providers and the challenges faced by the service providers in their bid to ensure adequate supply of water.

It is true that the available sources of water in most countries (that is the rivers, dams, springs, wells and rainwater) can serve many people well upon proper provision, distribution and management. Now, domestic demand is influenced by various factors among them; economic, social and technological, religious, geographical and water habits. These factors should be inherently considered to precisely estimate the water demand for a given community. Some countries receive abundant amount of rainfall resulting into a heavy surface runoff and flows in streams and rivers, but due to lack of storage facilities all this water is not harnessed for use when required. There is therefore the need to identify suitable dam sites to tap and store this water to meet the rapidly increasing domestic, industrial and irrigation water demand.

Some of the boreholes that are in existence in some places are not functional because of high operational costs simply because they are electrically powered, and cost of electricity is too steep. The possibility of using solar powered pumps should therefore, be explored. Their revival will contribute towards increasing the water coverage in the country. To further increase irrigated land in a country, dams and pans that have silted over time can be rehabilitated and water abstracted using appropriate low cost technology like solar power and the water used for drip irrigation. When some of these challenges are understood, then amicable solutions can be proposed to the various service providers in a country and ensure efficient supply and flow of water to the people.

CLEAN AND SAFE WATER PROVISION SERVICES

For people to access clean drinking water, water has to be supplied from its source to the consumer. A service is therefore required to provide this entity to the consumers without inconvenience. The water supply service providers are in charge with the overseeing water harvesting and collection. This includes among other things such as water storages, extraction of groundwater and replenishment of the same and also extraction of water from the rivers. They are also responsible for the transmission of water from its sources to the consumers. Other responsibilities include water reticulation. In other words, water service providers should oversee the drainage of water other than the storm water and will oversee the treatment of water and recycling of the same. This is to ensure that the water supplied to the consumer is free of disease causing organisms. There are several categories of water supply service providers in a country. These categories are discussed below.

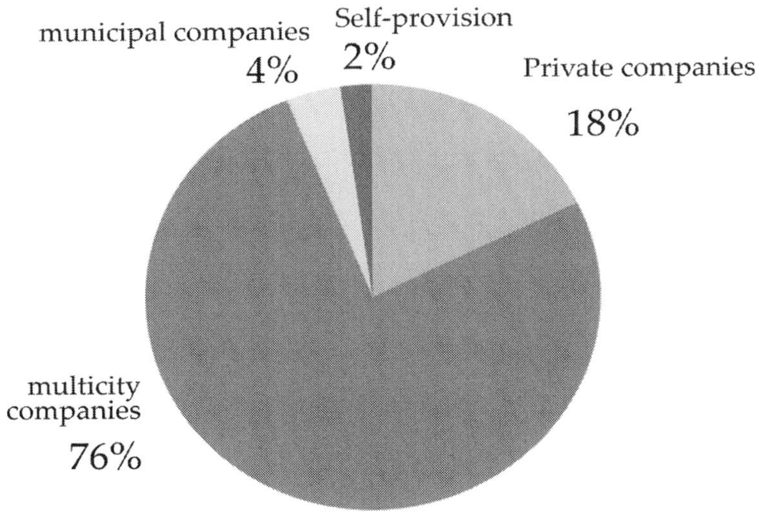

Source: http://sos.danubis.org/eng/country-notes/bulgaria/

Figure 1.3: Water supply service providers

Private sectors. More than 10% of water supplies in most parts of a country are provided and managed by private sectors or a collaborated effort between the private and public sectors. These entities usually provide water supply services under an agreement, a lease or a management contract. Under these arrangements, the public entity entrusted with the supply of water legally shall delegate some certain aspects of provision of the service to the private provider for a specified period of time. They account for a small percentage due to management and governance issues and hence most countries in the world have reverted to public water provision entities.

Public sectors. About 90% of water supply services in most countries around the world are provided and managed by public sectors. These public sectors are mainly owned by the local municipals, the state, cooperatives, or even collectives. They are nonprofit organizations with an aim of providing common benefit to the citizens. The challenge with this mode of service is that the inefficiency caused by mainly political interference leads to overstaffing and low labor productivity in most of the times. In the end, the losers are the poor who end up even paying more charge per liter of water than the rich. This is because the rich are well connected to water supply services utilities as opposed to the poor.

According to a statement given by Ryuto Hashimoto (former Japanese Prime minister) "Modest improvement of public water operators will have immense impact on global provision of services."

Faith-based organizations (FBOs). Different faith-based organisations have been playing a very huge role in providing and managing several water supply related services in most parts of the world, especially in developing countries. Some of the key responsibilities include financing water supply services and advocacy of water service provision.

WATER SUPPLY SERVICES PROVISION AND RELATED CHALLENGES

According to World Health Organisation (2015),

"Water is fundamental for life and health. The human right to water is indispensable for leading a healthy life in human dignity. It is a pre-requisite to the realization of all other human rights."

Provision of clean, adequate and safe accessible supply of water coupled with good sanitation programs is important in ensuring good health care. This will consequently reduce many of the diseases affecting the underprivileged people who either live in the slums or poor dry rural areas.

There are many factors that contribute to development of water supply services and provsion. Among these factors are time and energy saved during the search of clean and safe water for use.

A report by Water Supply Hygiene and Sanitation (WASH, 2005) showed the consequences of poor water supply coverage in most countries around the world is severe. Lack of access to clean and safe water causes higher infant mortality rates, low economic productivity, and low female enrolment in school. These consequences are more serious in the rural populations that have virtually no sanitation facilities.

According to WHO (2010), awareness trainings and workshops should be carried out in the communities on the need for the water projects. The community based organizations must also consider working under the management of the water companies to solve the challenge of management problems of water supply system.

Conducting a baseline water resource survey is of paramount importance in ensuring that there is adequate water supply in place for all. Inputs of experienced personnel in hydrogeology, geophysics, engineering, development planning and sociology are vital during water resource potential evaluation, appropriate and satisfactory site selection, and depth to groundwater establishment and choice of the right hand pump option. If evaluations such as groundwater resource potential and depth to groundwater are not well identified, the result would most likely be dry wells and, therefore, unsustainable schemes.

There are numerous challenges related to water supply and provision services. These challenges are discussed below.

Private and Public Sectors Challenges

Water scarcity. This is one of the challenges faced by most people around the world. For instance, the per capita fresh water availability in developing countries is 690 cubic meters per capita as compared to the global standard of 1000 cubic meters per capita. Some places are mainly arid and semi-arid, and this makes it challenging to provide and manage various water supply services because of the associated technologies and costs involved. It becomes more complicated in places where the effects of environmental degradation are compounded.

Very low or non-existent sewerage services. Another challenge in the provision of water supply services involves the lack of sewerage system and treatment plant in major towns and cities that deals with the waste water. Some places, especially big cities and towns in the world have poor sewerage plants connectivity. In anticipation of the increased urbanisation and population growth in most of these places, there is urgent an need to prepare and implement sewerage systems that are effective and efficient.

Poor water quality. This is another key issue challenging water supply services in most parts of the world. High levels of chemical concentration in ground water, both organic and inorganic is an alarming water quality issue that most people in some places must deal with. For instance, the very high concentration of arsenic in areas such as India, Nepal, Bangladesh, China as well as Taiwan and Malaysia makes the easy distribution of safe water impossible. Fluoride and Nitrate concentrations in fresh water sources are also a problem in waters around the globe. Oftentimes, this concentration usually goes beyond the accepted standards and makes utilization and use of water for various domestic purposes almost impossible.

Inadequate investment in water supply infrastructures. This is another challenge involved in water supply provision services. To achieve the Millennium Development Goals, water supply boards in respective countries require adequate "annual" capital for investment. However, the actual investments realized over the years have been way below in most countries, and this has continuously affected various water supply activities.

Non-revenue water supply services. The cash collected from the amount of water billed is much less than the amount of abstracted water. In some places this problem is usually heightened by dilapidated water supply mechanisms which lead to breakages and pipe burst. Also, most residents fail to pay for the water service provided forcing the water company to disconnect the links hence increasing the cost. Poor water distribution network leading to high water consumption that are unaccounted for, lack of consumer meters results in inaccurate billing and low revenue collection.

Inadequate source of water. In some places such as arid and semi-arid regions-where there is no definite rainfall pattern, water is not enough as different communities usually abstract water from the only nearby source available. During the dry spell, these communities depend on the water for irrigation and they usually face several problems when the water is not sufficient enough to sustain their daily usage.

Other challenges include lack of laboratory facilities for water quality monitoring and control and inadequate storage facilities making it difficult to store enough water for a continuous supply from the source to the consumers. For this reason, water companies are forced to supply water for specific days in a week to the residents. Additionally, inadequate transport facilities making delays when doing the maintenance works is another challenging issue in most places. There is also the problem of poor planning of the towns and residential plots making the companies are incapable of mapping the networks for the laying down of the water piping supply systems.

Faith-Based Organisations Challenges

There are several challenges that are usually experienced by the faith-based organizations in the bid to supply continuous supply of water to their respective service areas. These challenges include;

- Insufficient funds to upgrade their pipeline systems, consumer water meters and maintaining manpower and labour. This usually leads to increased inefficiency in operation.
- Old infrastructure-most of the projects and infrastructure since the time they are initiated and installed, have degraded, depreciated and grown old over the years of not updating, and are no longer efficient in water supply provision services.
- Increased cost of electricity and diesel-this is particularly to those service providers who use pumping as their means of supplying water to the consumers.
- Increased levels of Non-revenue water-this has been heightened by the inefficient old pipes which burst and leak occasionally, lack of master meters and individual meters to measure the amount of water consumed.
- Leadership challenges-most of these faith-based organizations and self-help groups are managed by the committee established among the locals of the community. Many haven't hired skilled labour that will enhance efficiency and service delivery. This has led to several stalled projects while others being are simply being abandoned by the government.
- Politics of the land-many of these projects are politicized making the success of the project, most of the time to be fully dependent on the politics of the area. This has led to many incomplete projects while others have stalled completely.

Every person has a right to clean and safe water in adequate quantities. This therefore requires the roles of the various stakeholders to be redefined so as to achieve this target. Despite the above challenges the responsible water boards have to take necessary steps to ensure efficient provision of water to the citizens of the country through the licensing of the various water supply provision services in the country.

AVAILABILITY AND DISTRIBUTION OF FRESHWATER

Accessibility to clean and safe water is determined by several factors. Among these factors, availability and physical location are the most important. Many researchers argue that timely availability, reliability, and accessibility to clean, safe, and adequate water plays a very significant role in satisfying various needs of humans. In other words, this ensures that there is water security in households for basic needs and economic activities. According to United Nations Environmental Programme (2015), the main sources of freshwater for socio-economic activities globally include rivers, pipe water, protected and unprotected springs, open wells, protected wells, roof catchments, dams, earth pans, boreholes, streams, lakes, reservoirs and underground aquifers resulting from a complex hydrological cycle.

Rivers. According to studies, most people especially those in developing countries, rely on rivers for water supply in their daily activities. The distribution of households based on source of water indicates that about 36% of households worldwide get water from rivers, 10% from pipe, 7 % from protected springs, 15% from unprotected springs, 2% from open wells, 1% from protected wells, 6% from roof catchments, 18% from dams and earth pans and 5% from boreholes. Households in less developed countries face a chance set of diverse sources which include house tap connections, public or private wells and taps, water vendors and tank trucks. Most rivers around the world are experiencing a decline in water flow and some are drying up in some parts due to indiscriminate deforestation of the catchment areas. The quality of water flowing in rivers has also been adversely affected through point sources of wastewater such as cattle dips, car washing, slaughter houses and municipal wastewater. Bathing in the river is very common in some parts and people carry household goods, clothes, and utensils to wash in rivers.

Rainwater. According to hydrological studies, it has been shown that the levels of precipitation in most parts of world have changed remarkably over time. This has contributed to huge precipitation variations between different seasons. In some places, water storage especially in the upper catchment area is usually endowed with plenty rainfall while the lower parts experiences arid and semi-arid conditions. This influences the availability of water from one place to another.

Dams, springs and boreholes. Dams, springs and boreholes are important sources of water for domestic use and in most cases, dams are silted and most of the springs are unprotected. The dams main purpose is to serve as storage or reservoir of water supply, while some communities also use it to generate electricity.

On the other hand, the filtration being naturally undergone by spring water removes the microbes from the water making is generally safe to drink. However, the issue becomes different when springs are used for community water supply. This time, the spring must be "protected" meaning, adequate monitoring and safety measures must be undertaken to ensure the maintenance of its safety.

The need for taking care of the water quality for boreholes or dug wells are the same as that of springs and dams. Though the use of boreholes, dug wells and tube wells are usually flexible-meaning it can be terminated when it is no longer needed, the frequency of sanitary inspection is as important as springs and dams, however, regular monitoring must be performed on boreholes, since ther are waters forcibly extracted from the water table. The tubes are basically sunk and extended way below during dry season to ensure continuous water supply.

The protection of the dams, springs and boreholes are also needed since the waters maybe contaminated as it emerges to the surface by contaminated surface water itself, wild and domestic animals and even people in contact with them.

Magnitude of Water Demand and Distribution

The demand and distribution of water in most places around the world includes the requirement for water services for domestic uses, agricultural activities, industrial activities, and other uses. Distribution of water can be defined as "delivery of water to customers with appropriate quality, quantity, and pressures."

Source: http://www.ec-dossenheim-zinsel.ac-strasbourg.fr/archives/2701

Figure 1.4: An illustration of the current situation of water accessibility in Sub-Saharan Africa (SSA)

Among the world's regions, Sub-Saharan Africa (SSA) faces the greatest problem of water scarcity because the freshwater resources are unevenly distributed over time and space. Africa uses only 4% of its water resources, yet it faces the greatest problem of water shortage. Despite the abundant fresh water resources in Africa's lakes and rivers, still a great disparity in water availability and use is experienced. Most freshwater in Africa comes from rain along the equator, specifically around the Congo basin and Niger delta. Saharan Africa and other third world countries experience several challenges. This ranges from the water infrastructure to the distance they have to walk to get this precious commodity.

The spatial demand and distribution of water resources vary depending on a number of factors such as willingness to pay, infrastructural capacity, modern technology, distribution systems and components.

Water availability is ensured by its reservoir; however, the distribution is a problem since water demands have increased over time while the supply system has lost its automatic function which has resulted in an on-request system. In most cases, water users have no fixed time schedules for water supply and therefore, people particularly in far and remote areas experience shortages especially in times of high demands. According to UNEP (2000), problems in distribution of water are caused by timing and scheduling and not by too small infrastructure capacities as the water users' claim. The deficiencies of the system lead to unclear, unequal and stressed water distribution. The basic of a modern distribution network is intended to meet the actual water demand, self-cleaning velocities of 0.4 m/s regularly in the network and the unidirectional flow.

The main characteristics of a modern distribution network are a branched system with pipes of relatively small diameter. In the design process, several steps can be recognized such as the determination of the water demand, arrange sections, compose main structure, design sections, check of pressure drops and fit in flows. The supply point is the 'end' of the distribution network as far as the water company is concerned. The water meter is the last part of the connection and in unmetered situations; the stopcock should be the last part of the connection. Then the actual house installation starts and that is where the water is consumed. For calculation reasons, the pressure at the actual supply point should be at least 200 kPa as the minimum acceptable pressure when no water is abstracted from the installation. Reason behind this is that the pressure at the highest tap point is enough to overcome the hydraulic resistance of the house installation and the tap point itself.

The amount of water needed for beneficial use should be calculated based on a simple per capita taking into account the projected needs of the population, industrial, commercial and other uses supplied by the permit applicant. According to UN-ESCAP (2000), water demands are estimated using population projections, land use information and water production records. Historical population trends and projected growth patterns provide a comprehensive insight into the future water demands determined by the magnitude, direction and characteristic of population growth. It is important to emphasize the need for an economic efficiency operational criteria requiring the distribution of stream flows until the net marginal utility in per capita water consumption is achieved. The relation between the quantity of water used and the price should be illustrated by the demand curve.

FACTORS DETERMINING RESIDENTIAL WATER DEMAND

Research studies show that the overall access to safe drinking water (both urban and rural) in the world has increased from the previous 41.2% in 2003 to 47.3% in 2005 and 2006. To cope with increased demand for water, most countries have tried to move towards the commercialization of water supply so that water supply services would reflect the true cost of supplying the service. This mainly requires the appropriate management of the demand for water services through pricing.

Towards the realization of commercialization of urban water supply systems, it is therefore necessary to have an in-depth understanding of the factors that affect the demand of water in a country. However, it appears that limited information exists on water demand in many parts of the world; to date, there is no such study undertaken either for large cities or for peri-urban towns. Consumption patterns for residential water in the peri-urban towns have changed over time due to the expansion of private tap connections, increases in population, expansion of house hold income generating activities that consumes water to a greater extent, improved sanitation practices and other shift variable variations.

Majority of people in most urban and peri-urban area are poor, with a monthly per-capita income of less than 100 US dollars. However, access to basic services such as drinking water, electricity, sanitation, as well as telecommunications, although limited, has improved over the last decade in most places around the world. Most people in various cities and towns rely on private piped connections, public stand pipes, private water vendors and open sources such as rivers.

According to World Health Organisation, there are frequent service scrambles (a service scramble means that water is made unavailable to some households for a period of piped water because the number of currently connected households is more than the expected capacity of water source, due to a rapid increase in the number of people in various cities and towns around the world.

In order to alleviate this frequent problem of service scramble, the water supply companies in collaboration with funding from the World Bank's Urban Water Supply and Sanitation Project in most countries are undertaking the expansion of the capacity of water supply system and distribution lines to ensure that there is adequate water supply connection for all. Some of the major factors that determine residential water demand are discussed below.

Housing and Other Related Characteristics

Households who own a house are more likely to obtain water from taps in their residence because the directives of water supply office in most places around the world usually require tenure certification as a perquisite to obtain the permission and the service of piped water. It might be also because of the length of time of stay at the current home, as households who own the house are more likely not having tenure problems and thus are supposed to stay at the house for indefinite amount of time in the future. This convinces them to have a private pipe connection whereas households who do not own the house where they live are uncertain to invest in a property that does not belong to them. Most of them have a mindset that private pipe connections are more of the responsibility of the landlords.

Households with higher monthly expenditure are also more likely to obtain water primarily from private pipes. This is probably since better-off households are likely to be able to pay the service and connection charges required in advance by the water supply office, hence, expenditure is considered as a proxy for household income and livelihood.

Similarly, educational status is also found to affect the decision of households to have private piped connections. This could be related with the better awareness and understanding of educated households on the positive returns in improved health and sanitation associated with piped water.

Factors like the size of the house and access to appliances such as showers, bathrooms, washing machines, flush toilets, and availability of garden or lawn in the residence influence water demand. It is expected that housing ownership is likely to increase household's decision to have private piped connections as the primary source of water and is likely to increase the daily per capita water consumption. Thus, for both water source choice model and water demand estimation model, housing ownership is expected to have positive sign.

Water Pricing

When a price of water exists, then it is quite easy to compute a per unit price for each household and each source. The price of non-piped water has been considered exogenous in most research studies that has been done previously. But there are other researchers that argue that the price of vended water is endogenous because price is determined by demand and supply factors. Due to the fragmented nature of the water vending market, household decisions of water demand are likely to influence its price. Even if free of charge, the collection of water from non-piped sources usually involves costs for hauling water from distant sources.

The rate at which users pay for water bought from public stand pipes, water vendors or other sources is different from the rate of private connection or from free sources that have opportunity costs. This would create inconsistency if similar price is included in the analysis. The demand of water in most parts of the world is mainly perceived as inelastic in its nature. This is because there is no a perfect replacement of water basic usage. In addition, most people do exhibit low levels of perceptions on water rating because bills for water supply usually represent a very small proportion of their income. This is especially evident in developing countries where significant portions of people, who make decisions of expenditure, are not yet fully aware and open to paying for water supply or reluctant to know the rate structures.

Social-Income Status

Statistically, there are key factors related to social-income status that have significant impact on the daily per capita water consumption. These are primary employment of household head, expenditure, age and sex of the household.

From these factors that are usually perceived to affect the quantity of residential water demand of households, monthly expenditure of the family and primary employment of household heads are positively related with the dependent variable, and daily per capita water consumption. Male headed households and age of the household head in most cases usually have negative impacts on daily per capita water consumption.

Monthly expenditure also have a positive relation with daily per capita water consumption of a household because family members of better households are more likely to have frequent bath and showering, frequent washing of clothes and more water for washing dishes and cooking water as compared with those households that needs to control water consumption due to financial limitations, hence, taking into consideration the household lifestyles and sanitation preferences of each types of households.

DETERMINING ACTUAL WATER DEMAND

Per Capita Approach

When using per capita approach in determining the actual water demand of an area or a place, the first stage involves the determination of the present "per capita use" of water supply system in that area. This should be expressed in terms of cubic meter per capita. For instance, if you are asked to determine the actual water demand between the 2005 and 2010, there are several steps that should be followed, and each step should be used to determine a specific number which is then tabulated in a spread sheet. These steps are outlined below:

- Determination of the total amount of water that was withdrawn in 2005 in cubic meters (annual average). This amount can be expressed as "WD2005"
- Determination of the total amount of additional water that was derived or bought from other sources in 2005, and this should be expressed in cubic meters. This amount can be expressed as "PD2005"
- Determination of the total amount of water that was sold to other water supply systems, and should be expressed in cubic meters. This can be referred to as "SD2005"
- Calculation of the total amount of water in the supply system, which is expressed as "TSW2005". This amount is determined by getting the sum of WD2005+PD2005 minus SD2005.
- Determination of the total population expressed as (POP2005) which is serviced by the water supply system within the area in 2005.

- Calculation of the "Per Capita Usage" for the year 2005, which is expressed as (PCU2005). This is obtained by dividing the total water in the supply system by the total population serviced that year. That is

$$PCU2005 = \frac{TSW2005}{POP2005}$$

- Determination of the actual number for the year 2010 using the steps from 1 to 5.
- Calculation of the "Average Per Capita Use" (**PCU base**) for 2005 and 2010. This is done by adding them and then dividing the result by two. That is

$$\mathbf{PCU\,base} = \frac{PCU2005 + PCU2010}{2}$$

This represents the present "per capita use" of water supply system.

Goodrich Approach

This formula is mainly used to estimate the water peak. In other words, it is used to determine the average water rate on a daily basis, and this is done based on a specific timeframe. This is usually expressed as:

$$P = 180.t^{-0.10}$$

Where,

P = represents the average annual rate in percentage (that is volume/day) used in a specific timeframe

t = represents the time in which the demand peak of the water is required (valid period of time usually range between 2 hours to 360 days)

The q √ n Approach

Another approach is the "q root n" method used to determine maximum water demand for a number of houses and is adequate for the design of modern distribution networks. This is expressed as:

$$qM_n = o.083\sqrt{n \times TU\ houses}$$

Where,

qM_n = represents the optimum instantaneous demand of n houses in *l/s*

n = represents the total number of houses

TU = represents total number of tapping units house and

0.083 = represents the capacity of 1 TU in *1/s* (300 *1/h*)

The demand pattern of one house is much more erratically than the demand pattern of 1,000 houses. The pattern of one house will be several spikes over the day, while the demand pattern of a cluster of houses will have a much smoother pattern. The method to determine the maximum demand for several houses is based on the method used to determine the maximum demand for an in house installation.

Water Demand Function Approach

This is mainly used to represents the single demand of water that is provided from a tap. This approach usually assumes that there is no any other alternative commodity that can be used to substitute water. This approach is mainly used as a function for households water consumption rate (q), water price dependant variable (p) and demand shifter vector (x)- for example, characteristics of the household, weather conditions, and house equipment mainly to regulate the preference and other variables that may affect the demand of water in a household. A water demand function formula is expressed as below.

$$q = f\left(p, x\right) + u$$

Where,

q = represents the quantity of water consumed

f = represents the function

p = represents the price

x = represents the demand shifter

u = represents the error term

The error term u is added to this relationship to account for unobservable or measurement error in variables.

EFFECTIVENESS OF WATER USE MANAGEMENT STRATEGIES

There is the need to develop and implement water use management strategies which should be aimed at influencing the distribution and demand of water in a certain area. This helps to attain adequate level of water utilization and maximum consumption of the finite resource. There are several strategies that can be adopted to achieve optimal water management in a community. These strategies are discussed below.

Water pricing. The price signals provide the information needed to allocate and distribute water more effectively. The water pricing is a major instrument that can be used to resolve inevitable trade-offs. It forms part of the broad effort that strengthens water use governance and value long-term ecosystem services. Water pricing ensure that there is efficient water resource utilization with an aim of achieving a high level of water cost recovery. Water conservation measures can be promoted by structuring the entire water rating system and charges to express the various variations of water usage, distribution of water especially during the peak hours, and the production cost of the water.

Water metering. This is another very important and most effective strategy that is used in many places for efficiency and optimal conservation of water resources. Water metering is required for water loss control, accounting and rate making. Inaccurate reading may lead to inaccurate information about water consumption and leak detection. Water metering strategy can be used to help in detecting and controlling the amount of water demand and waste.

The main aim of water metering strategy is to try and manage in the optimum, any water loss within the economically acceptable levels of usage. Where a water supply is fully metered, the consumption patterns of various categories of consumers can be determined by monitoring the meter readings over a period of time. All water suppliers experience water losses through leakage in the transmission systems, illegal connections, faulty meters and wastage. In general, this is estimated at 20% of the total supply but poorly managed systems may experience losses at even greater than 50%.

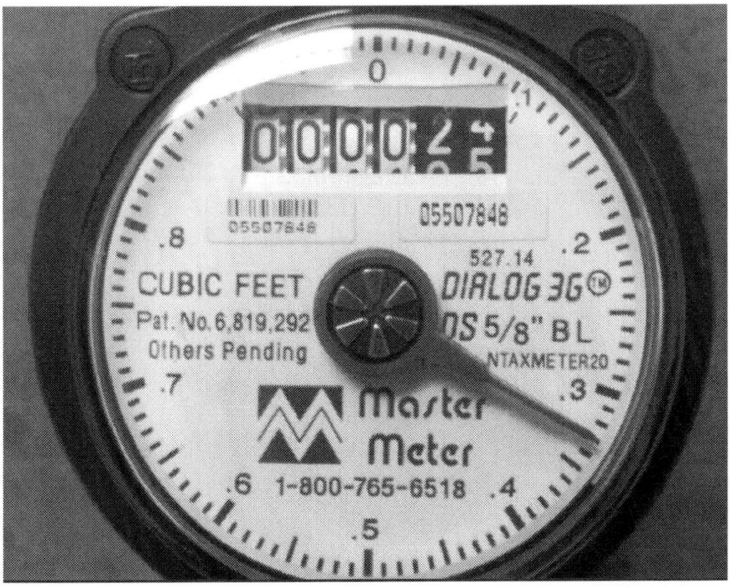

Source: https://earthzine.org/2015/07/07/meters-matter-how-water-metering-promotes-conservation-and-technological-development/

Figure 1.5: A typical water meter used in water supply system used to measure actual water consumption

Rationing/restriction and prohibitions. Water use restrictions and prohibitions are part of command and control tools. Water regulatory agencies and water service providers establish restrictions and bans in water use during droughts for non-essential purposes such as washing cars and watering flowers. Water service providers use water rationing as a tool to manage water use. Those served by water supply companies note that "they have to discipline themselves because they are usually aware that water will only be available to them at certain times, therefore, they must use water for the basic needs only and this reduces wastage". Rationing is set for effective supply of the demanded volume of water to every zone. For example, low lift water pumps receive 50 cubic meters in 24 hours while high lift receives 75 cubic meters, 11 hours per night.

Public information initiatives. Water customers should understand the water bill which clearly identifies charges, rates, and any other information on the pertinent water usage such as measures taken to conserve water at household level. The bills on water consumption can be used to provide essential information of water supplying companies such as contact information that can be used to contact the providers in case there is a problem with water supply network such as leakage and water wastage.

Sensitization on effective use of water. Workshops in schools and churches by water officials are the major means of public education that can be used in any area. Such sensitization events can also be done by water supply staffs at market places in their respective areas of operation. Among the topics that can be covered in the public education forums are the basics of water conservation and why do it, water use management practices like reporting leaking taps, burst water pipes, water recycling and control of free flow of water from taps.

Water reuse and recycling. Greywater (wastewater) that comes from various source points can be treated to remove all the foreign materials and then reused for other specific functions such as flushing the toilet and redistributing for irrigation. This strategy can be used to reduce water use, mainly through efficiency and effective conservation mechanisms.

2 EVALUATION OF PIPED WATER SUPPLY SYSTEM AND GROUND WATER QUALITY

PIPED WATER SUPPLY SYSTEMS

Worldwide, systems in water supply are the most advanced, successful, and accomplished services in the history and their existence have greatly impacted positively on the lives and welfare of most people. In developing countries, urban areas are growing rapidly. The expansion is not only affecting the city planning but also the urban water in the aspect of inadequate quantity of water supply. Increase population, especially in the urban areas has led to a rapid increase in water demand and put more pressure on the pipe infrastructure.

Most countries use pipe water system as the main water distribution network of clean potable water however this has always posed different challenges to the water supply in many countries in terms of accessibility and sustainability to the increasing water demand of the fast-growing population. Some of the challenges that are faced with pipe water supply system include expansion of pipe, rehabilitation in case of damage, and replacing the entire system whenever need arises.

Source: https://thefunambulist.net/architectural-projects/india-water-infrastructure

Figure 2.1: A typical pipe water supply system

Utility service providers which are owned by the government are responsible for water supply. Over the years the pipe supply system has become insufficient to meet the growing consumer water demand. The main concerns are with the water pipe system management especially with new installations having to go through bureaucratic steps to be approved, maintained and replaced. The increasing water demand would not be a problem if the government had enough resources but the hiring of workforce for pipe management and ordering for new pipes are constrained by lack of funds.

The increase in population in most cities is an additional constraint to the pipe water supply system as it has resulted in water supply system being placed under a lot of pressure as the available pipes in place are unable to sustain the growing population. In addition, the total pipe network length in place is not enough to serve the increasing water demand from the population.

The average water supply in most places, especially in developing countries has been growing at an average rate of 6% for the past few years. This has caused the water accessibility to increase, but not fast enough to cover the increasing consumer water demand causing a gap between the water demand and water supply.

VARIABLES OF PIPED WATER SUPPLY SYSTEM

The figure below indicates the primary variables that influence water supply systems in a place. These include, assumed (exogenous), endogenous and the excluded (ignored) factors.

Exogenous are profiles that originates from the community itself. It can include but not limited to the data on births, deaths, the current water tariff being imposed in an area, the net migration over a period and adjustment times. This are sometimes attributed to consumer-related factors that are essential in the design of the piped water supply system.

Endogenous factors are the internal factors that directly affect the piped water supply system. This covers the vast operations, the management and logistics concerning pipe water supply system. This includes the water supply itself, the water demand, income, expenditures, available funds, workforce, population, pipe integrity and the desired water tariff which must be correlated with the exogenous factors to make the pipe water supply system work.

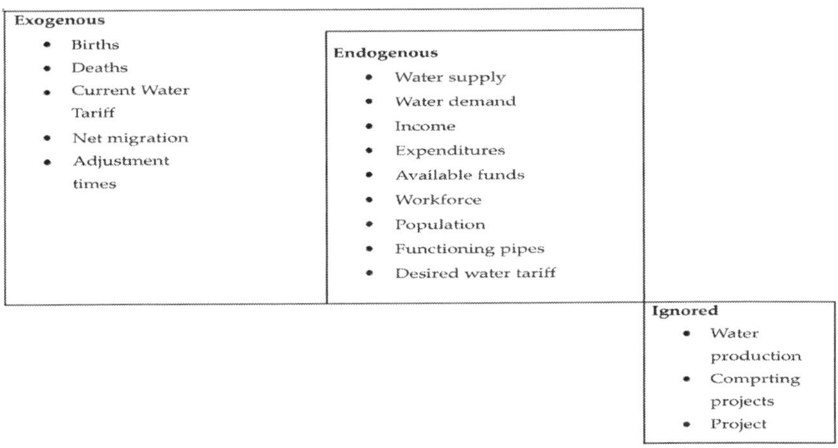

Source: http://rsta.royalsocietypublishing.org/content/371/2002/20120417

Figure 2.2: Shows exogenous, endogenous and ignored variables in piped water system

Ignored factors are not literally ignored, but these factors are the ones that weigh the least over the three. Though water production is surprisingly under the ignored factor, this is mainly because that water production remains a fixed factor that needs to be custom-fitted with the existing exogenous and endogenous influences. This factor does not fluctuate most of the time and readily determined. The same goes for competing projects and other projects. These factors are already existing and its contribution to piped water system is already anticipated and determined and not much of a change as compared to the first two factors cited.

While based on the above factors discussed, Figure 2.3 below identified the three sectors that must be emphasized in the water supply system. The first sector is the population sector which determines water demand. The second one is Pipe sector (water supply system) which structures the way the water board manages the pipes in the system, from the moment they are reported to when they are ordered and replaced as well as how the pipe supply system is controlled. The third sector is the Finance Sector. In this sector both major sources of income and expenditures should be structured. The figure below is a model illustrating three key sectors that interact with each other in water supply system.

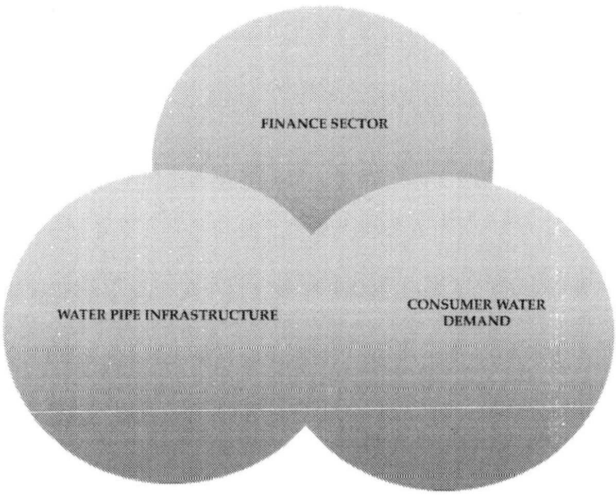

Source: http://www.scielo.org.za/scielo.php?script=sci_arttext&pid =S1816-79502013000200016

Figure 2.3: The relationship between the different sectors in water supply system

The main role of the model is to understand what causes the limited accessibility of potable drinking water that results into a water supply gap in place. The water supply gap is the difference between the desired water demand and actual water supply.

The total population of a place and the average water domestic consumption determines the desired demand per capita (liters/week/person). On the other hand, the water supply defined by the functioning pipes, average water domestic consumption and pipe density (pipe density is the total number of people supplied per pipe).

The hypothesized reason why there is a water supply gap is because of poor and ineffective pipe water management system presently in place. The explanatory model aims to assist in identifying the different existing dynamic systems that could be behind the ineffective pipe system management.

The model works in a way that there are already a certain number of functioning pipes in place. These are the pipes supplying water to the entire population of a place or country. These pipes are maintained on a regular basis (Regular maintenance stock). There is also a probability that these pipes could get damaged or obsolete (Damaged and obsolete pipe stocks). The obsolete pipes can also be defined as depreciated pipes but they are not damaged but just old. For instance, the average life time of pipes used in an area is 20 years and after this life time these pipes are expected to be replaced but sometimes some of the pipes get damaged while awaiting to be replaced, which calls for obsolete damaged pipes emergency replacement rate. These pipes should be accumulated into the obsolete damaged stock.

Finance Sector

There should bc a continuous weekly pipes maintenance and replacement due to pipe breakages and leakages, which is a very expensive activity. In addition, due to the growing water demand from the increasing population, the pipe system must be expanded. This implies that the management water board needs to buy new pipes and hire a workforce for the installation and maintenance of the pipes. The finance sector in the model helps to understand how much funds are available. The reliable funds come from within the water pipes system itself so if the system is poorly managed, the water board management won't have much funds especially if the entire pipe system breakdown and is not replaced in time. The income is determined by how much revenue of water is obtainable in relation to expenditure (actual water supplied minus water lost), so if there are too many pipes not replaced, income is going to be low because water supply will be low.

The expenditures in the system are total sum of investment costs in new pipes, total workers costs and weekly maintenance costs. With the finance in place, the management is able to know how much workforce can be hired and how many new pipes can be ordered. So in a nutshell, the funds have a great impact on the pipe water supply effective management.

The damaged and old pipes are supposed to be replaced and new pipes must be installed for expansion planning as regards to the growing population to keep the water supply system balanced. In order to do this, a pipe supply control system is needed.

In the pipe supply control system new pipes can be ordered. In order to do so, there is the need to know how many pipes are needed in the system (total pipes renewal and installation demand). This demand for the pipes is determined by capacity demand for the pipes to be replaced and added (obsolete, obsolete damaged, damaged, planned pipeline expansion pipes). Each of these stocks has its capacity demand determined by using the accumulated stock divided by the normal repair time. The repair times differ from each stock pipe.

With the total demand determined, the pipes are ordered but there is a constraint faced from the funds at times (effect of available funds on ordering rate). The ordered pipes take a substantial period of time to be delivered to the inventory stock.

For pipes to be finally installed there is the need for a workforce in place, the hiring rate for this workforce depends on thc total pipes demand and workforce productivity estimated at (3pipes/person per week), and of course the funds needed to pay the workers (workforce budget). Pipe installation also depends on pipe installation capacity.

Finally, pipes are installed according to urgency. The management needs to know how much capacity is available for each stock that has pipes for installation or replacement. This is done by getting capacity residual minus available capacity for the stock. Capacity residual is what is left after a certain capacity has been deducted. The management replaces the damaged pipes first, obsolete damaged pipes, new pipes for pipeline expansion and lastly obsolete pipes.

Water Demand (Population Sector)

The Population sector in the model is associated with the water demand in a place. The dynamics behind the population include births, deaths and migration as seen and explained below.

The birthrate is expressed as the area population ratio of the fertile women population (this in other words means fertile rate over fertile period. Women fertile rate (i.e. the average children born per woman) is calculated over time and the fertile period is 35 years (the time a woman can stay fertile). The rate is divided by 52 to determine births per week.

Net migration rate that also increases the population can be determined using the equation:

$$Net\ Migration = \frac{Area\ population \times urbanisation\ fraction}{52}$$

The urbanization fraction is assumed to be 0.028 and the rate is divided by 52 to determine the migration increase per week. The population is decreased by deaths. Which is usually expressed as:

$$Death = \frac{Area\ population \times Death\ fraction}{52}$$

Death fraction is usually assumed to be 0.035. The rate is divided by 52 to determine deaths per week, while water demand is determined to be:

$$Water\ Demand = Average\ Water\ Domestic\ Consumption\ per\ Week \times Area\ Population$$

The domestic consumption is how much water each person consumes per week (liters per person per week) which is estimated at 29 liters.

Pipe Sector (Water Supply Sector)

The pipe sector is currently the main water source distribution for potable water in most places around the world. The Pipe sector is divided into two parts, (that is pipe supply control system and the functioning pipes management system).

The pipe sector focuses on management and efficiency of pipes in a water supply system of an area. Therefore, the main stock in this sector are the functioning Pipes, which are currently the pipes in the system.

The total number of the Functioning Pipes can be determined using the population served with water /pipe density.

There are two ways of increasing pipes in the water supply system. The first step is through installation of new pipes using the expansion pipeline or through repair, maintenance and replacement of pipes that are damaged or obsolete. All these pipes are accumulated in a stock called Functioning Pipes. The ways of increasing pipes in the water supply systems are detailed as:

Through new installations. When there is an increase in the consumer demand, the water management is obliged to expand the water system to meet the demand. A planned pipeline stock in this system helps in the regulation of the functioning pipes. The stock is increased by the Pipeline Expansion Planning Rate, which is determined by the formula:

$$Pipeline\,Expansion\,Planning\,Rate = \frac{Desired\,Number\,of\,New\,Pipes}{Adjustment\,Time\,For\,New\,Pipes}$$

This is a nonlinear equation that helps to determine how many new pipes are added per week over a period. In other words,

$$Desired\,Number\,Of\,New\,Pipes = Desired\,Number\,Of\,Total\,Pipes - Pipes\,In\,Place$$

The equation determines the gap between total required pipes and how many pipes are actually available. The desired number of total pipes is derived from area Population/Pipe density where pipe density is the number of persons per pipe. Pipes in place are all the pipes in the water supply system. In this case a summation of all the stock pipes should be done under an assumption that all pipes old and damaged need to be replaced at some given point of time. If the damaged pipes and obsolete pipes are not included there will be double counting because the pipes in need of replacement will be considered in the pipeline Expansion Rate and later will also be replaced after regular replacement times-including too many pipes in the system and resulting in heavy expenditure on maintenance. The planned pipeline expansion stock is decreased by the Additional Functioning Pipeline Installation rate which at the same time increases the functioning pipe stock. This rate is determined by:

*Installation rate = (Available Capacity for New Pipes) *Effect of Workforce Ratio on Installation Rate*

This effect is captured by using a graphical function because under the assumption that it is in a non-linear relationship. The equation works in a way that if the workforce is a high, more pipes are installed and vice versa.

The second way of increasing pipes in the water supply system is through *repair, replacement and maintenance of pipes*. Functioning pipe to be managed efficiently have to go through regular maintenance thus the need for the maintenance stock.

All old and broken pipes must be replaced. Different stocks categorize the repairs and replacements under the following classes; obsolete pipes, Obsolete Damaged Pipes and Damaged Pipes. For each of the replacement rate, each stock capacity demand is determined by dividing the stock by the repair time. But First, the total demand has to be determined which leads to the pipe supply control sector.

The pipe supply control sector is responsible for pipes orders and inventory. The Pipe ordering rate accumulates the pipes on order stock and it is determined by the indicated orders:

Equation = (MAX (0, indicated orders))*Effect of Available Funds on Ordering Rate.

A "max" function in this case is used so that the rate can pick the highest number as the indicated demand, however, if the number is negative, the rate takes 0 as the indicated demand. Once the pipes are ordered its takes 6 weeks for them to be delivered to the pipes on inventory stock. The pipes delivery rate determines how many pipes are on order divided by the time it will take to be delivered. That is:

$$Pipes\ Delivery\ Rate = \frac{Pipes\ on\ order}{Delivery\ time}$$

The pipes on order stocks and pipe on inventory stock should both be adjusted over a period of time- approximately 18 weeks.

With pipes on Inventory, the Indicated Pipes Usage (how pipes are used from the inventory) is the number of pipes available to be installed, and this depends on the normal inventory coverage- that is the number of weeks it normally takes to receive the inventory divided by pipes on inventory.

A workforce is needed to install or replace the pipes and the target workforce is determined by:

$$\textit{Target Workforce} = \frac{\textit{Target Production per Week}}{\textit{Reference Workforce Productivity}}$$

Where,

Target production represents the total pipes renewal and installation demands, whereas reference workforce productivity represents the number of pipes installed or repaired pipes per worker per week. The Workforce Net Hiring Rate should be adjusted over a period of 12 weeks and the rate is constrained by the availability of funds.

AN ASSESSMENT OF GROUND WATER QUALITY

Water is a basic human right and to achieve this right, the water sector players require not only to increase access to water but also to manage the water quality. Development of water resources allows a country to realize this constitutional requirement. According to Water Supply and Collaboration Council (2010), the problem of access to safe water in the developing countries is enormous and therefore to alleviate the problem, governments in these countries have a responsibility to give priorities to implementation of community water projects.

It has been argued several times, that water is essential for life and all human activities must have water source; "*it may be dirty, it may be inadequate in volume and it may be several hours walk away but nonetheless some water must be available.*" Different types of water supply systems are appropriate in different areas. Water supply systems are projects implemented essentially for providing water for either industrial, domestic or animal use or one of the three purposes. They are dependent mainly on the available sources. In general, water projects can be grouped into three main classifications which includes:

- Classification by location (urban and rural water project)
- Classification by ownership (private or community water project)

- Classification by the source of water (rain water harvesting, surface water, or ground water project)

The quality of water supply to rural communities is critical in developing countries. According to the Water Supply and Collaboration Council (2010), almost 900 million people worldwide use water from unimproved sources, a threat to their health and productivity.

Safe drinking water that is fresh and free from harmful chemicals and disease-causing organisms therefore is a pre-requisite for a healthy population.

Human developmental activities among them-agriculture, urban settlement and industry interfere with the quality of water from various sources. Water is finite and therefore can easily be contaminated or depleted by human activities, causing water stress regions of the world. It has been revealed that among various causes of preventable mortality and morbidity in the world are poor sanitation, poor water quality, as well as poor and inefficient supply of clean and safe water.

Water quality varies from region to region depending on climate and geology. For instance, boreholes in high rainfall areas produce relatively fresh water compared to those that are found in areas with low rainfall. Some of the boreholes in semi-arid areas produce highly mineralized and saline water.

Ground Water Quality in Relation to Geology

Ground water is an important resource in most places around the world. Ground water is used for various purposes among them domestic water needs and small scale irrigation. The kinds and concentration rates of different materials in groundwater are dependent on rocks which are in contact with water. It has been argued that sedimentary formation with respect to igneous and metamorphic rocks plays a major role in water quality. Therefore, geological formation is one of the most important and effective agents of the quality of water. For instance, Miocene and Marl formation affect the groundwater quality and increases its electrical conductivity. According to the kind of petrology of each formation, water may affect different geological formations and dissolve different cations and anions, which in turn affect the quality of water in these aquifers. Therefore, the sedimentology of quaternary rock formations can determine the hydrology and hydrogeology of water of a specific aquifer. In a study conducted in Iran, it was found out that the quality of groundwater was changed by the underlying geologic formations.

There are several parameters that may be associated with groundwater quality. Some of these parameters include pH, colour, turbidity, conductivity, iron, total hardness, total alkalinity, fluoride, and total dissolved solids. Rainfall does not only affect the groundwater quality but it also influences the quality of groundwater. Total hardness in water may be due to magnesium and/ or calcium salts. It has been observed that increase in total hardness in surface water can be attributed to the decrease in the rate of evaporation at high temperature. On the other hand, groundwater salinity increases during the dry season where no precipitation event occurs.

Table 2.1: World Health Organisation (WHO) Drinking Water Quality standards

ELEMENT	UNITS	WHO STANDARDS
pH	pH Scale	6.5-8.5
Colour	mgPt/l	Max 15
Turbidity	N.T.U	Max 5
Conductivity	(25DegC) uS/cm	Max 2500
Iron	mg/l	Max 0.3
Manganese	mg/l	Max 0.1
Calcium	mg/l	Max 100
Magnesium	mg/l	Max 100
Sodium	mg/l	Max 200
Potassium	mg/l	Max 50
Total Hardness	mgCaCO3/l	Max 500
Total Alkalinity	mgCaCO3/l	Max 500
Chloride	mg/l	Max 250
Fluoride	mg/l	Max 1.5
Nitrate	mgN/l	Max 10
Nitrite	mgN/l	Max 0.1
Sulphate	mg/l	Max 450
Free Carbon Dioxide	mg/l	-

Total Dissolved Solids	mg/l	Max 1500
Arsenic	ug/l	Max 10

Source: http://pubs.sciepub.com/ajwr/4/5/2/index.html

Variables Influencing the Quality of Groundwater

Groundwater quality influences sustainable water utilization as well as safe water availability. Figure 2.4 below shows the variable relationship between the groundwater quality and the various factors that affect it. Among them; climate, geology, hydrology, water abstraction and land use.

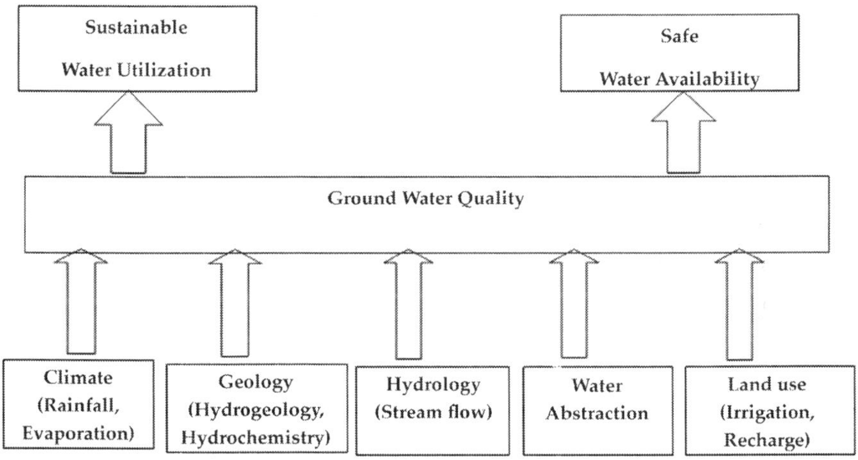

Source: https://www.slideshare.net/AtunRoyChoudhury/ground-water-sampling-analysis-technique

Figure 2.4: Schematic diagram showing variables that influence the quality of ground water

Spatial variations on water quality. Studies have indicated that there are always significant spatial variations in the groundwater quality parameters in most places around the world. The variations of the different water quality parameters are usually characterized by increase or decrease in independent directions. The spatial variations do not seem to be interconnected. The causes of the spatial variations are mainly attributed to the nature of geological formations of a place.

Temporal variations of ground water quality parameters. Both short term and long-term changes in the ground water chemical parameters are usually found to exist in most places around the world. The Electrical conductivity, Total Dissolved Solids (TDS) and fluoride are usually found to increase when baseline conditions are considered. The values for TDS in some cases are usually above the WHO standards although all the values may remain intact. The increases and or decreases of these parameters are usually attributed to the nature of the geological materials within the specific aquifers.

The influence of hydrogeology on ground water quality. The ground water quality variations are usually attributed to the differences in the composition of the geologic formations at the different aquifer levels. The influence of hydrogeology on the ground water quality plays a key role in the variations since other factors including rainfall are significant in influencing water quality changes.

The influence of water abstraction and land use. Research studies have indicated that there is no significant influence of the water abstraction on the ground water quality. However, it should be noted that there exist a strong relationship between abstraction rates in the dry seasons and ground water quality than in the case during the rain seasons. It was also found out that the daily activities practised by residents within an area (such as farming, use of fertilizers, and other chemical) generally influence the ground water quality in that area.

The levels of key water quality parameters. Most of the ground water quality parameters such as turbidity may be within the WHO standards. However, some of the parameters such as electrical conductivity, TDS and fluoride may increase and are likely to surpass the WHO standards at any time.

CONSERVATION AND EFFICIENT USE OF WATER

In cities across the globe, stormwater runoff causes pollution to enter the stormwater sewer system and is often discharged without treatment into local bodies of water. Some major pollutants of water include pet waste, litter, motor oil, cleaning products, paint, construction debris, and outdoor usage of pesticides. Flooding increases the magnitude of this problem as it causes a large amount of water runoff that contains waste. Illegal connections to the stormwater sewer system are another problem that complicates the situation. These connections may contain an illicit discharge that is, a dischargc other than stormwater, and the extra water in the sewers increases the chances of flooding. The severity of water pollution throughout the world is often overlooked or understated. In a 2005 report, the National Environmental Education and Training Foundation found that 78% of the American public does not understand that runoff from roads, lawns, and agricultural land is currently the most common source of water pollution. Also, 47% of Americans believe industry is still the most significant contributor to water pollution. In general, public awareness needs to improve in regard to the contamination of the water supply and as well as of nearby bodies of water.

Most places around the world have been facing several concerns with waste water management system including flooding and polluted stormwater runoff into local bodies of water. Despite the many permits given to different boards to monitor and enforce legal and regulate various illegal activities that threaten the quality of water in different countries. There is still a significant level of negligence among different governmental boards in their responsibilities in governing and conserving the quality of water bodies in respective countries around the world. One of the most prominent problems experienced in most places around the world is the lack of public awareness and lack of behavioural change to prevent water pollution from occurring.

A good example of encouraging behavioural change includes informal education. For instance, a city like New York has embraced an educational campaign that focuses mainly on the impact of polluted stormwater discharges in local bodies of water and the various steps citizens can take to mitigate the amount of pollution.

Source: http://www.latimes.com/tn-dpt-me-0326-water-education-20150325-story.html

Figure 2.5: Educational water campaign with kids as participants.

Raising public awareness through educational campaigns may be the only feasible solution to the problem of water pollution because of how everyday actions cause the pollution. Yet there needs to be an effective method to measure the success of how well an educational campaign is reaching out as well as its impact on the ways in which people behave. Those community members who are already aware of the significant problem of polluted water unfortunately are not educated about their own negative contribution. Also, individuals are not being given a specific set of management practices to follow. Every city that has created a successful public education program about polluted stormwater runoff should reach out to a range of audiences including businesses, residents, landscapers, and tourists. If public awareness is only spread to neighbourhood residents and does not extend to businesses and tourists, the campaign will not effectively reduce stormwater runoff pollution. Moreover, people need to be informed of how the polluted water can negatively affect them to realize the urgency of the problem.

The United States Environmental Protection Agency proposed the utilization of the Best Management Practices (BMPs) to address problems of water pollution. BMPs can range from educational programs to physical devices put in placc to correct the flow of water. This chapter contains information about different EPA guidelines that a country must follow to regulate their sewer systems.

WATER POLLUTION

Pollution causes problems for the environment and the government. Every day people take to the streets in cars, motorcycles, trucks, and SUVs. Carbon emissions from burning fossil fuels pollute the air on a regular basis. Fouling the air directly harms people by affecting their lungs; however, it also has consequences that are not as evident to the average person such as ozone damage and greenhouse gas build-up.

Another area of the environment that is impacted by human activities is the water. Governments design systems to send clean water into pipes and pump the water into buildings to be used for multiple purposes. Cities and towns provide treatment for waste water so that it can be recycled. When expected circumstances change, the water treatment system will not work properly. Bacteria, trace metals, and pharmaceuticals are just a few of the pollutants that can pass through the system untreated. Flaws and oversights in the purification process allow these harmful pollutants to stay in the water after it has been treated.

In 2005, an extensive study was conducted in London to evaluate the levels of pharmaceuticals in its drinking water system because of the failure to properly treat the wastewater. The levels of pharmaceuticals in the water varied seasonally and varied with the sewage treatment facility that was studied. This problem has only recently come to the forefront of pollution research. Modern sewage treatment plants do not screen or treat the water for pharmaceutical drugs. When humans take medicine, birth control pills, or weight training supplements, they increase the body's chemical levels. The body then tries to correct its internal levels and many of the increased hormones or antibiotics are flushed out of the system through the urine. The urine makes its way to the water treatment plant where the chemical drug pollutants pass through the system untreated. Finally, hormones and chemicals end up in bays or similar bodies of water and cause unnoticeable effects on the aquatic life as well as to humans. This "soft" pollution is not pollution in the traditional sense of the word.

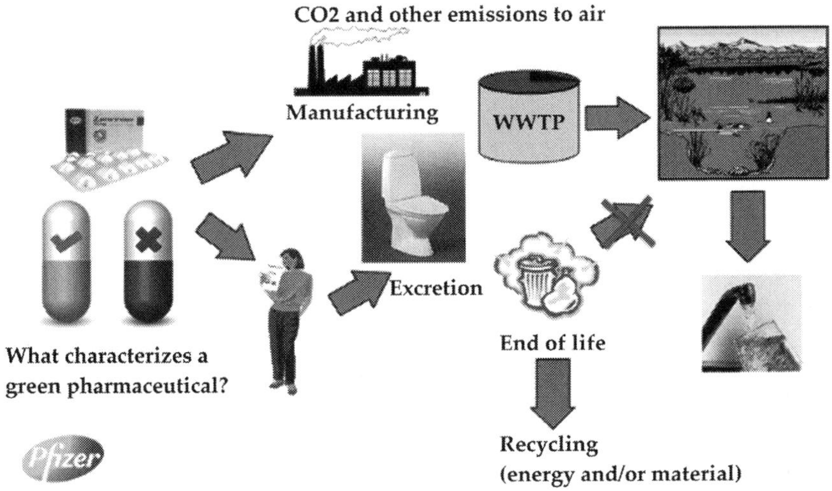

Source: http://www.ansvarsblogg.se/miljo/green-chemistry-lecture-at-ki/

Figure 2.6: Illustrating how pharmaceutical products end up in our water supply system

The water does not look any dirtier, and the water may even taste the same. However, just because the person or animal consuming the water cannot detect anything does not mean pollution is not present. For example, birth control medications are having negative effects on aquatic wildlife. The chemical ethinylestradiol, which has been classified as an Endocrine Disrupting Substance (EDS) has been attributed to the feminization of fish. EDS are substances that affects a living organism by disturbing or interrupting the functionality of the endocrine glands. The endocrine glands is responsible for the secretion of chemicals, particularly hormones for male and female. Different sets of substances are being releases based on the gender. For the aquatic wildlife, which are relatively small, the concentration of ethinylestradiol in the water has the capability to interrupt fish endocrine functions, yielding errors in hormonal release. In other words, by the presence of these EDS a male fish can develop an ovary-like structures in their testes, or vice versa. The levels of these chemicals in today's waters are not enough to affect humans, but when left unchecked and present in drinking waters, these chemicals becomes a possible threat to the systemic functions of the human bodoy. There are other forms of water pollutions as discussed below.

Stormwater Runoff and Combined Sewage Overflows. Stormwater runoff contributes greatly to the pollution of the water in and around large cities. Combined sewage overflows (CSOs) greatly increase pollution when they become overwhelmed because they drain directly into bodies of water. Normally, water goes through a water treatment plant first and after proper treatment and purification the water is discharged into a nearby body of water. Unfortunately, during times of extreme rain or melting of snow, the system can become overwhelmed by the influx of the additional water. During these situations, the system is designed to drain the water directly into the nearby bodies of water. This means that all the sewage, waste, and pollutants picked up on the streets flow directly into a clean supply of untreated water. For instance, combined sewage overflows from a country like India contribute 27 billion gallons or more of raw sewage and polluted water into clean water bodies. Therefore, direct action to correct or shut down combined sewage overflows would significantly decrease direct contamination of natural bodies of water close to large urban areas.

Illegal Connections. Illegal connections and illegal discharges affect the usefulness of the stormwater system. They can also put the water supply system at great risk of filling up and overflowing onto the streets. These illegal connections that hook up to the system without a permit cause major pollution.

According to United States Environmental Protection Agency, construction sites, gas stations, car washes, restaurants and barber shops are examples of potential or actual polluters of water resources. The government and other responsible agencies should focus on the BMPs in apprehending and managing water pollutants to prevent and control the menace. BMPs does not implement any end-of-pipe pollution limitation systems, rather, it focuses on simple activities such as street sweeping, catch basin maintenance, and identification of illicit connections to the storm drains. This may help in preventing stormwater contaminants from entering the stormwater system and creating illicit discharges.

Animal Waste & Flooding. Research into the impact of animal waste as a source of pollution has not been widely conducted. Animal waste is a problem is basically existing and has been an emerging cause of water pollution for decades around the world.

According to an agricultural census conducted in 2007 by United States Department of Agriculture (USDA), about 2.2 billion of livestock (cattle, swine, and poultry) generated around 1.1 billion tonnes of waste (manure) in 2007. It was stated that if these waste is not managed and maintained appropriately, various pollutants (such as bacteria, ammonia, and nutrients) found in the animal manure may end up degrading the quality of surface water and ground water. This will basically endanger and post threat to human lives and ecological health.

Source: http://www.pewtrusts.org/en/about/news-room/press-releases/2013/06/05/pew-disappointed-in-epa-plan-to-study-impact-of-industrial-livestock-operations-on-chesapeake-bay

Figure 2.7: A cattle farm that generates animal waste which is a source of water pollution

Many people neglect collecting and disposing the waste of their animal in a well-maintained manner. In other words, no one would pick up the waste left behind by their animals. They might bag it, but then they would have nowhere to put the waste, and so they leave it in the park.

According to a study done in Lagos, Nigeria, animal waste is the single biggest component of litter by weight. It contributes about 1,200 tons of waste each year.

On the other hand, flooding is a huge issue that leads to the pollution of the streams, bays, and lakes in many places around the world. It not only increases pollutants, but also affects the lives of citizens living in flood prone areas. Traffic can shut down due to a submerged section of road or underpass. Commutes that normally take ten minutes can be extended to hours. Flooding in cities also leads to direct flow of contaminated water into local bodies of water from non-point sources. Instead of flowing into the water through one location, it flows along the streets and into the water from a multitude of locations and across a wide area. Flash floods can claim lives as well. One second a road will be perfectly navigable and the next the influx of stormwater fills the system, and it overflows onto the street. The water system could also overflow because of blocked drains due to poor maintenance of the city's basic infrastructure. Therefore, Best Management Practices (BMPs) can be adopted in order to help in correcting the problem of sewers overflowing. Best Management Practices range from physical attempts to correct the stormwater system or prevent failure of the current system, to public awareness campaigns. Physical solutions could be increasing the capacity of the system such that when the storms hit and place huge demands on it, the system can handle the influx of water.

Source: http://www.euronews.com/tag/china-flood

Figure 2.8: Example of a flooding event in an urban area

Other solutions may include placing devices around storm drains to prevent clogging or blocking. In areas where debris blocks the access of water, these devices would allow the system to continue to function properly and divert the flow of water off the streets and into the sewer or stormwater pipes. Finally, public awareness campaigns help prevent humans from aggravating the shortcomings of the system. Litter, pet waste, car maintenance, and failing septic systems can lead to pollution of the water and overloading the storm water pipes. Getting the public to focus on these issues and presenting simple solutions can be a very effective way to eliminate problems in the system.

For instance, a country like China has put in place numerous flood prevention and flood mitigation techniques. China deals with the issue of inner city flooding every year especially when there is heavy down pour during monsoon seasons. The country mapped the municipal flood plain by identifying structures above the flood plain and those in danger below the flood plain. Then, it used government money to elevate structures in the areas with the greatest risk. This preventative measure was the one most extensive step taken. These measures were mainly preventative and can help to decrease pollution by removing utilities that could spill contaminants into the water.

WATER POLLUTION MITIGATION

EPA Regulations

Not all regions and sewer pipes in the United States are regulated by the EPA. There are some urban areas that fit into the EPA's Phase I while others fit to the Phase II. All Phase I includes large or medium cities or areas with populations above 100,000. All those citizens in this phase are required under EPA regulation guideline to apply for permits for coverage of their stormwater discharge. On the other hand, Phase II requires that small urbanized areas and their immediate surroundings apply for permits to cover their stormwater discharges. Other than the areas under these phases, the stormwater flow remains pretty much unregulated. The locales listed in Phase I and Phase II are required to create comprehensive stormwater management programs (SWMPs) to help in reducing contamination and prohibit illicit discharges. Moreover,

Phase I and Phase II regulations do not pertain to combined sewers. They only regulate municipal waste separate storm sewer systems (MS4s), which are defined by four criteria.

- The pipes are designed to handle stormwater
- The pipes are not for combined sewage and stormwater
- MS4s are owned by a city or state, not a federally owned sewage treatment facility
- MS4s empty their contents into U.S. waters.

Several United States Environmental Protection Agency mandates to address the management of MS4s. One is the Total Maximum Daily Limit (TMDL). The EPA standardized the units for waste as pounds/day of phosphorous for waste load allocations (WLAs) or point sources. Municipal stormwater sources are regulated as point sources and as such must document their discharges in pounds/day of phosphorous. The load allocations (LAs) are pollutants from nonpoint locations or urban runoff, improper animal handling, forestry practices, or failing septic tanks. The amount of pollutants a municipality can discharge into the bodies of water depends on which body of water they they use as the recipient. The size of the body of water, the previous level of contamination, and the rate that fresh water mixes with the polluted water are all factors that determine the TMDL for a city. The TMDL is calculated using the equation:

$$\mathrm{TMDL} = \sum \mathrm{WLA} + \sum \mathrm{LA} + \mathrm{MOS}$$

Where,

WLAs = represents the point sources

LAs = represents the nonpoint sources

MOS = represents the margin of safety

By adding a margin of safety, any huge influxes of pollutions due to storms or natural disasters such as fires will not cause the level of pollution in the waters to become toxic. As long as cities follow the TMDL, their waters should remain safe, but the samples that cities use to test their water are not always correct or accurate.

The EPA's (2011) Clean Water Act and the National Pollutant Discharge Elimination System (NPDES) are two sets of regulations that every country around the world needs to follow. The Clean Water Act makes it unlawful to discharge pollutants from a point source unless a permit has been obtained. The NPDES permits allow for cities to dump pollutants into bays and other U.S. waters. Private homeowners do not fall under the scope of this legislation. This bill pertains to companies, governments, and facilities that discharge pollutants directly into bodies of water. The Clean Water Act also established other duties for the United States Environmental Protection Agency. It established standards for regulation of basic pollutants and gave the EPA the right to put into place programs that would control and decrease the amount of pollution. Also, the Clean Water Act funds the creation of sewage treatment plants that can handle the new amounts of sewage being sent in from local businesses.

Best Management Practices (BMPs) Recommendations

It is believed that if all people around the world were to focus on best management practices, it will help mitigate storm water pollution. Best management practices should be tailored towards urban residents while other should be focussed towards people in the community. Regarding businesses (such as restaurants), the use of grease traps must be enforced as well as the proper procedure its disposal. The person who picks up the grease from the grease trap must have the appropriate certification to ensure that the grease is being disposed properly. Also, restaurants should receive a receipt that states the date and time of the disposal in case an inspection should occur.

Another BMP that can be recommended in terms of water pollution control and prevention measures is the use of a training program that should be carried out by the responsible authorities such as the utility companies and municipalities. Rather than initially fining businesses or restaurants of incorrect practices, it would be appropriate that the violators be required to attend a training program to educate them on the important things that they need to know as businessmen with respect to proper waste handling and disposal. This can be done through an educational campaign presentation. If, after completing the training program the business or restaurant still does not cooperate or change its habits, then a fine may be imposed.

An important BMP that can be used in residential, hotel, and commercial areas is the use of biodegradable soaps. When washing sidewalks, walkways, or entrances, the use of anything besides biodegradable soaps should be considered. The use of biodegradable soaps would prevent any illicit discharges from entering the stormwater drains.

Poorly kept garbage areas also contribute to stormwater pollution. A BMP that can be recommended would be to require that all buildings put their garbage in a consolidated, blocked off area where stormwater could not invade. It would be appropriate to construct a lift or elevation in the ground that surrounds the entire garbage area so that runoff cannot sweep under garbage areas or dumpsters and pick up potential debris. Another recommendation would be to construct a drain that would be connected to the regular sewer system, which could be especially beneficial to restaurants and hotels. To prevent debris from being carried away in runoffs it would be recommended to use grass filters. This would be very beneficial for individuals or businesses when washing their driveways, sidewalks, or building entrances.

In particular, a pump should be used to properly dispose any runoff and deliver it to the proper sewage drain in low-lying areas where dumpsters and garbage are present. A BMP that can be recommended with regards to this concern is to require certification of the use of this pump and its connection to the standard sewer system (as opposed to the stormwater sewer system) to ensure that these garbage areas are being kept well and that runoffs do not enter the stormwater system.

According to the principles of BMP, all water pipes should be installed underneath sidewalks rather than having them empty out over the sidewalk. If a pipe empties out clean water over the sidewalk, it would potentially pick up debris and residue from the sidewalk that then ends up in the stormwater drain. By having pipes go underneath the sidewalk and directly into the street, it decreases the chance of picking up trash or anything that could contaminate the water in general.

To mitigate stormwater pollution on a broader scale, another BMP that can be emphasized would be to install grates in the stormwater system that would not allow any debris to filter through into the pipes. This would be a multi-level sewer in which the stormwater would filter through the grate while any debris or trash picked up along the way would not enter the system.

Public Awareness Campaigns

In addition to limiting the levels of pollutants in the water, the United States Environmental Protection Agency believes in implementing public outreach programs. The programs address the behaviours of the community that cause pollution of water. Littering, not properly cleaning up animal waste, and ignoring guidelines for disposal of controlled substances are a few elements that the EPA addresses when trying to change the public's behaviour. The United States Environmental Protection Agency focuses mainly on positive reinforcement and increased awareness. It attributes most of the poor behaviour of the public to ignorance of the greater consequences of small actions. By presenting facts that link small, seemingly harmless actions with larger negative consequences, people are able to see the bigger picture. Also, convenience is an issue for many residents. When listing off solutions that solve a problem, residents must understand that they will not be inconvenienced by potential remedies to water pollution problems. Indeed, a solution that is too complex or requires people to go out of their way will likely be rejected by the public or ignored after only a few attempts. Many times, the best solutions are the simplest.

Educational campaign should focus on the impact of stormwater discharges on water bodies, the pollutants of concern and their sources, and steps the contributors of stormwater and non-stormwater discharges can take to reduce the pollutants. Some of the examples of educational campaigns include refrigerator magnets, utility bill inserts, and even trainings for businesses on new techniques for reducing pollution.

3 IMPROVING WATER SECURITY IN ARID AND SEMI-ARID AREAS THROUGH RAINWATER HARVESTING

INTRODUCTION

Water is a crucial element for human survival; however, the predominant challenge is the increasing freshwater scarcity given that the projected worldwide water demand exceeds supplies. Some of the key regions that safe or are vulnerable to water scarcity are the Asia, some parts of South America, and Horn of Africa (including Ethiopia, Eritrea, Sudan, Somalia, and Kenya)

Some of these places have been experiencing decline in water availability per capita over time. For instance, the current per capita in a country such as Ethiopia is 650 cubic metres from 1800 cubic meter in 1960. By the year 2020, these levels are predicted to even fall to as low as 350 cubic meters per capita. There are several causes of these declines. Some of these causes include reduced precipitation brought about by climate change coupled with environmental degradation triggered by human encroachment on watersheds, and hydrological variability and trans-boundary issues surrounding water resources, as Ethiopia and other adversely affected areas shares some of its water sources (rivers, lakes and aquifers) with her neighbouring countries. As a result, the country experiences food crises and resource conflicts due to competition over dwindling supplies of water, pasture and arable land.

Different countries have introduced different measures in order to help in curbing issues to do with water insecurity. One of the key measures entails reform on water governance institutions through the Water Act which provides for public-private partnerships to allow alternative service providers to supply water to unreached populations such as slum and rural dwellers. Such alternative service providers included small companies, vendors and community based organizations. Although any water act is supposed to revolutionize the water sector, much focus should be laid on surface and underground water development. However, the desired levels of universal access to water for all have not been achieved, especially in countries experiencing arid and semi-arid climatic conditions.

Regardless of the deteriorating water situation, some countries in the world continue to rely heavily on surface water and ground water to meet their daily water needs even though records indicate that surface water and groundwater are being over-extracted. Nevertheless, these countries receive as far as 500mm mean annual rainfall which, according to various climatologists, if sufficiently harvested and stored above the current levels which fall below 150 cubic meters, can increase water availability to address the water insecurities in the marginal lands. It is evident to state that insufficient rainwater harvesting is responsible for extreme water scarcity in arid and semi-arid areas around the world. It is certainly enigmatic for many in a country to suffer water shortage while run-off, often causing destruction to property and lives, is not captured for use.

Such is the outcry for a shift of approach to water security in arid and semi-arid areas of a country. For instance, a country like Sudan usually receive heavy rains around the month of April to the end of June, but due to failure to harvest rainwater adequately, the country has persistent water shortages, especially on the northern parts.

Rainwater harvesting is a technique that every country should embrace to enhance water security in arid and semi- arid areas of a country. Therefore, it is paramount for a country to investigate the causes of failure of rainwater harvesting to achieve its potential and improve practices, policies and institutions that could encourage effective rainwater harvesting techniques so that there is more reproducibility, scaling up, and uptake, of effective rain water harvesting in arid and semi-arid areas.

RAINWATER HARVESTING POTENTIALS IN ARID AND SEMI-ARID AREA

Most arid and semi-arid areas around the world have the potential to be water sufficient and address the water insecurities. For instance, for rural households, an area receiving just 300 mm annual rainfall has as much potential for roof water harvesting as one receiving 2,000 mm. Assuming the per capita rural water consumption at 20 litres per day shows an annual water demand of 7.3 cubic metres per person per year, which could be supplied by a roof catchment of 36.5 squared meter, if 300 mm of rainfall per annum were available. A country like Ethiopia has a rainwater potential through domestic roof catchment of more than 450 cubic meters. If harvested and managed, this water is enough to support more than 250 million people or close to three times the current population of Ethiopia. However, it is vital to bear in mind that rainwater potential constitutes more complexities and these potentials depends on several factors other than the catchment size, which include , the average annual rainfall, the catchment type, the population density (for roof catchment) and slope gradient. Nevertheless, this rainwater potential has not been tapped and a great proportion of rainwater in most parts of the world drains away as run-off.

Unfortunately, there is lack of satisfactory statistics on the rainwater use in different countries even though rainwater harvesting is not a new technology to many people worldwide. Based on this argument, it is possible to assume that most areas around the world, with diminishing springs and underground water sources, and largely underserved by piped water could be depending on mostly rainwater.

This hypothesis addresses the question of why rainwater has not been harnessed to levels that could minimize water shortage in arid and semi-arid areas. According to the UNEP (2005), lack of adequate rain water harvesting policies, institutions and governance framework could be responsible for the dismal exploitation of rainwater in a place. A number of climatologists have stated that lack of rainwater harvesting specific policy and operational guidelines could be partly responsible for the poor rainwater harvesting status.

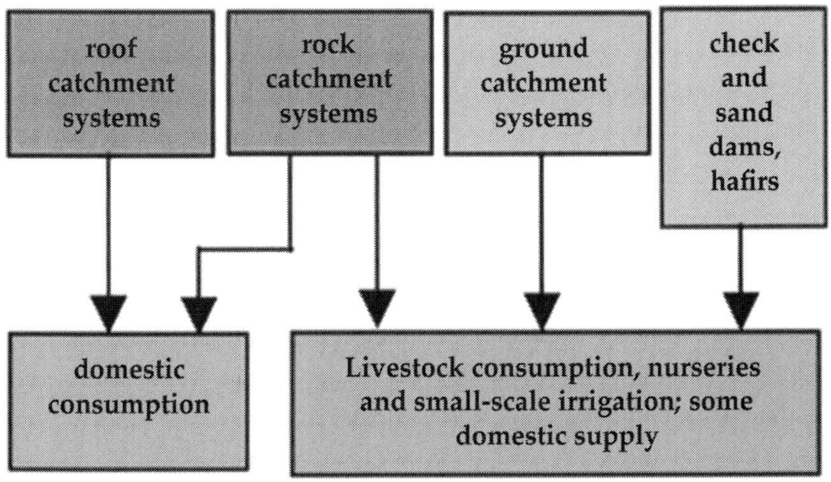

Source: http://www.lboro.ac.uk/well/resources/fact-sheets/fact-sheets-htm/drh.htm

Figure 3.1. Types of rainwater harvesting techniques

Most countries have not accorded rainwater harvesting as a priority in their strategic planning, and this has significantly limited to and dependent on special government or donor funds to enhance water supply systems in different areas of a country.

According to a survey that was conducted by UNEP in the year 2005 in sub-Saharan countries, there are different successes that may be achieved through the practice of domestic rainwater harvesting. However, a major mechanism for legal and policy issues for the mainstreaming and integration of rainwater harvesting in public water supply is missing in most sub-Saharan countries (which are mainly arid and semi-arid). This agency further stated that knowledge of the water sector organizations about the spread of rainwater harvesting is usually weak and there is little experience in establishing legal instruments to promote that spread.

One of the key barriers on rainwater harvesting in most places is priority articulation within the national and at county level water strategies. It has been noted that in many place, rainwater harvesting has often being implemented outside the legal water resources management framework, on a project level, which therefore, obscure many limitations because, while individual village-level projects succeed in the short term, their long terms sustainability could be severely tested in the absence of an appropriate institutional and legal framework at all levels. Moreover, it is also unlikely that extensive replication of appropriate technologies and community-based implementation strategies would be achievable in the absence of supportive institutions at higher levels, even if isolated project success at the local level may be possible.

Lack of supportive policies to guide how lessons learned are applied and how further investments could be directed to promote growth are fundamental and urgently needed in the rainwater harvesting sector. For a country to be water secure, it has to embrace coordination and integration of rainwater harvesting activities in mainstream water management plans. Most rainwater collection in different place is practiced haphazardly with little or no linkage and coordination among projects and among actors, which, in most cases, usually lead to duplication of efforts and little or no replication. Additionally, because of poorly designed rainwater harvesting and management systems, much of the water is usually wasted through evaporation and percolation.

In summary, it is the responsibility of specific governmental and non-governmental boards to try and investigate various practices of rainwater harvesting, various enabling and constraining elements and to ascertain or identify any new institutional arrangements that might promote more effective rainwater harvesting management in different arid and semi-arid areas.

TECHNOLOGIES IN RAINWATER HARVESTING

Most rainwater harvesting projects are usually implemented in schools, communities and households. Normally, the rainwater harvesting systems are composed of three main components, namely;

- The rainwater catchment subsystem
- The water conveyance subsystem (which include gutter and pipeline)
- The storage subsystem

A roof (made of galvanized steel sheet) for instance, comprises a catchment surface while the gutters and pipelines are used for conveying runoff from the roof into the storage usually in the form of tanks (can be plastic tanks, underground reservoirs, Ferro-cement tanks).

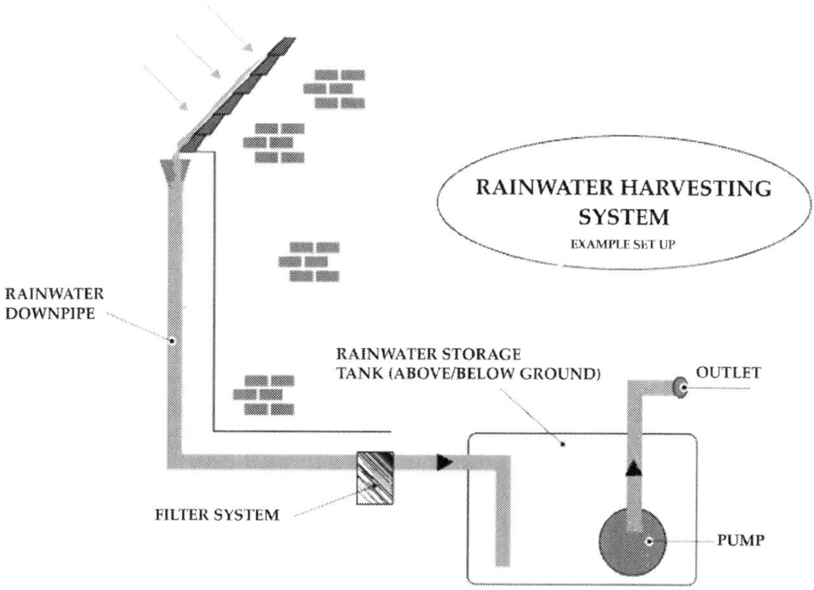

Source: http://www.rainwindsun.com/faqs/rainwater-harvesting.html

Figure 3.2: A simple rainwater harvesting system

There are several factors that influence the classification of a rainwater harvesting system. Some of these factors include the source of the run-off and the type of storage method used. In most places, runoff water is usually collected from the ground catchment and from the short transitory streams (such as footpath drainage systems, roadside drainage system, and floodwater).

Other examples of catchment include natural surfaces, gullies, rills, croplands, pasture, roofs and hill slopes. The storage is either in separate structures (ex-situ methods such as tanks, reservoirs, dams and water pans) or soil profile (in–situ methods like terraces, etc.). Examples of ex-situ storage structures include underground water tanks composed of sunken pits lined with ultraviolet resistant polythene lining. Others include rock catchments, sand dams and subsurface dams (barriers constructed across sandy riverbeds, especially in arid and semi-arid areas).

Other than the two factors that are mentioned here above, there are others that influence the techniques used in harvesting rainwater in a place. These factors include geophysical characteristics like rainfall amount, soil texture, nature of the landscape and intended use of the collected water. Some examples of the various types of technologies categorized in two main types of rain water harvesting systems, ex-situ and in-situ, are explained and presented in figure below.

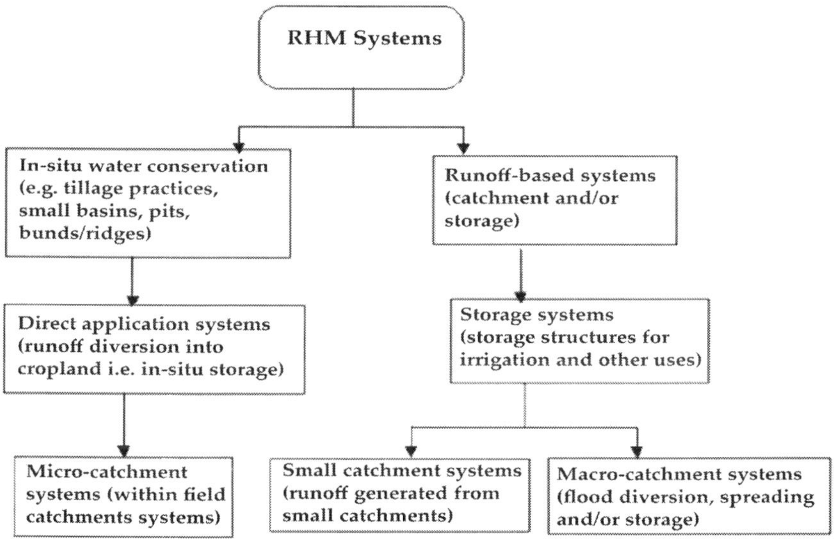

Source: http://academicjournals.org/journal/IJWREE/article-full-text/785D07451159

Figure 3.3: Rainwater Harvesting Technologies

Ex-situ rainwater harvesting technology: this involves catching rainwater mainly form the rooftops of building and then storing that water in either below ground tanks or above the ground tanks. Studies have revealed that a place with annual rainfall potential of more than 150 mm should be considered for roof catchment technology. This means that most areas around the world (arid and semi-arid) have high potential of rainwater harvesting which range from 0-200 mm per year. The type and size of roof (related to socio-cultural factors) are limitations. For instance, various pastoralist communities mostly reside in temporary structures whose mud roofs are not very conducive to roof catchment.

Surface rainwater runoffs are mainly harvesting from open areas and this water is then stored in different ground-based storage structures such as underground tanks, ponds, dams, water pans, and rock catchment. This type of rainwater harvesting technique is usually viable anywhere if geophysical conditions permit. Some of these geophysical conditions include soil texture which would determine whether a farm pond is lined or not, slope gradient which would affect location of dams, presence of sandy river beds for sand dams, rock out crops for rock catchment and amount of rainfall.

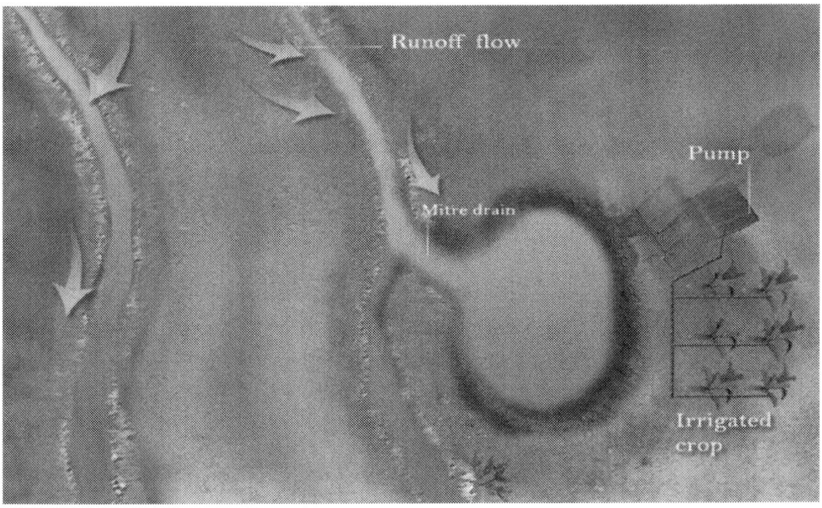

Source:https://www.researchgate.net/publication/265842519_Rainwater_Harvesting_Inventory_of_Kenya_An_overview_of_techniques_sustainability_factor_and_stakeholders

Figure 3.4: Trapezoidal bunds capturing runoff from a macro-catchment area

According to UNEP (2005), a typical household would need about and 400,000 litres for domestic, livestock and farming per a year. To capture 400,000 litres of rainwater would need a large roof surface area and several tanks which would be costly. It is therefore important to determine the type of rainwater storage facility that should be used before initiating the harvesting process. For instance, ponds can be used to hold larger volumes of water for a household at a lesser cost than tanks. Otherwise both ponds and roof catchment tanks can be applied so that the tanks supply water for domestic purposes while the ponds are utilized for farming and livestock needs.

In-situ rainwater harvesting technology. One difference between in-situ rainwater harvesting technology and the ex-situ rainwater harvesting technology is that this technology is used to retain much of the rain water right on the point it falls on the ground (for instance in pastured and cropped areas).

Source: https://www.theguardian.com/global-development/gallery/2011/sep/13/kenya-small-farmers-development-in-pictures

Figure 3.5: A typical in-situ rainwater harvesting technology

EVALUATING RAINWATER PROSPECTIVE IN AN AREA

According to World Health Organization (2010), evaluating the demand of water can be done using simple calculations based on water needs per use. For example, if the water use per capita is estimated at 15 litres per day and a typical household is taken to consist of six members, for a season (182.5days) the water requirement would be 16,400 litres (6people x15 litres x182.5days). Water requirement for thirty-five animals for a season would be estimated at 90,000 litres, while for farming on a quarter acre the water need would be 91,250 litres. The total estimated water demand for a household would be estimated at 202,940 litres (202.9 cubic meters). By applying these evaluation, contractors can determine the capacity of structures based on the number of households in the locality.

When determining the capacity of a rainwater storage facility (such as open reservoirs like dams) it is significant to consider various aspects of rainwater such as loss of water through seepage (which in most cases is estimated at 25% of the water in the reservoir) and evaporation (estimated at 50% of the water in the reservoir). Other aspects that should be considered include the texture of soil and whether the location is shielded or not.

Even though currently there are no integrated national statistics that could give an elaborate picture of the rainwater potential in most places around the world , researchers have been making effort towards this goal with the aim of generating data for enhancing advocacy for rainwater harvesting. For instance, a country like Ethiopia has an estimated potential for rainwater harvesting through roof catchment of more than 350 billion cubic meters. Another good example is a city like Nairobi in Kenya which has a mean annual rainfall of 1000 mm and thus a potential of 596, 485 cubic meters/day assuming 30% rainwater is captured. In other words, this water can be used to cater for more than 85% of water needs if 30% of rainwater was captured.

However, the amount of roof top catchment would depend on population density. An area with high population density would have more catchment areas due to more structural roofs than in sparsely populated areas.

IMPORTANCE OF RAINWATER HARVESTING

Different researchers and scholars have occasionally stated that effective rainwater harvesting may play a very critical role in enhancing water security and resilience to the various challenges experienced in a place due to climate change, especially for the communities living arid and semi-arid areas. It has been recognized that surface water and groundwater are usually depleted very fast since, they are usually over-extracted and it is therefore significant to emphasize the idea of embracing various technologies of water harvesting. This will help in reducing over reliance on various water resources such as rivers, streams, and aquifers. In doing so, it is important to increase the use of various techniques for rainwater harvesting such as diversions for runoffs, ponds, and rooftop catchment areas.

Rainwater has for a long time been considered cheap as compared to piped water, it is also non- exhaustible and can help to improve water stress among the most affected especially in the arid and semi- arid places.

A country like Kenya developed a Water Sector Strategic Plan 2010-2015. In this plan, it was stated that the country needs to reflect a major paradigm shift in its focus from extraction and distribution of water to overall objective of replenishment, conservation and development of water resources. This would depend largely on how water capture and storage is managed. Therefore, application of rainwater harvesting technologies would avert water crises by reducing overreliance on the streams hence easing pressure on the stream flow of the entire watershed.

Extreme climate accelerates the degradation of environment through increasing heavy flooding and prolonged droughts. The results are loss of property and lives, siltation of dams, rationing of water and power and increased difficulty in environmental restoration such as reforestation. Continued worsening of the country's water shortage situation is considered as water scarce and it is usually expected as climate change affects overall precipitation patterns.

Since economies of most countries, especially arid and semi-arid places are dependent on agriculture which is mostly rain-fed, reduced and sporadic precipitation affect both crop and livestock production and consequently leads to disintegration of livelihoods especially in the moisture stressed arid and semi-arid lands which are considered vulnerable and therefore more susceptible to impacts of climate change.

To address vulnerabilities associated with impacts of climate change on the water sector and strengthen community resilience and adaptation, appropriate measures such as rainwater harvesting techniques are needed to improve the community adaptive capacity. This is especially critical for communities in marginal lands which are considered very vulnerable to climatic changes.

Resilience, on the other hand is defined as "the ability of a system and its component parts to anticipate, absorb, accommodate, or recover from the effects of a potentially hazardous event in a timely and efficient manner, ensuring the preservation, restoration, or improvement of its essential basic structures and functions."

Rainwater harvesting techniques are listed among the specific adaptation measures that the water sector in most parts of the world need to be undertaken to cope with future climate change because of its high potential for addressing the impacts of climatic change on water security. However, there has been little progress towards achieving this in different places. In addition, approval and implementation of national policies and strategies on a country's rainwater harvesting techniques is also required in order to increase application of rainwater harvesting and storage in a country. Studies have also indicated that rainwater harvesting is one of the most appropriate and inexpensive technologies to reduce the level of vulnerability to water scarcity shocks and enhance sustainable livelihoods in moisture stressed ecosystems.

MANAGEMENT AND GOVERNANCE IN RAINWATER HARVESTING TECHNOLOGY

In different countries, the construction and implementation of rainwater harvesting projects is mainly funded by international organizations, enterprises and individuals. Although recently rainwater harvesting projects in developing countries such as Ethiopia and Kenya have received funds from international aid and cooperation programs as well as reserves from government budget of consolidated funds.

Private: In some instances, implementation of rainwater harvesting projects is usually private in the sense that the project is owned by a household. This is usually the case for projects like roof catchment, and farm ponds whereby individuals install rainwater harvesting projects in their homesteads, with or without assistance from donors, for their own use and with or without selling surplus water to the community members.

Communal: This is where a project is implemented by the community jointly. In most places, this is usually the structure utilized by many donors and government-funded projects. In such cases, land for installing rainwater harvesting project is usually provided by the community or a community member. The community usually nominate representatives to a management committee to oversee the project in trust. However, this is considered as a setback when projects are implemented on private land without implementers proceeding to seek change of ownership. Later this would affect sustainability or expected outcomes as owners of the land may end up claiming ownership of the project. Earth dams, sand dams, and subsurface dams and in some cases farm ponds are usually implemented this way.

Communal implementation of rainwater harvesting projects depends on the donor approach. Some donors would hold consultations with the community and facilitate formation of management committees, whereas others would survey existing/ registered groups and choose the one to partner to implement the project. The outcomes from the two approaches may be influenced by perceptions and the direct beneficiaries are usually the group members.

Combination of communal and private: According to this model, each member is supposed to contribute a given amount of money in order to get installation of rainwater harvesting and storage facility such as farm pond in his or her compound. The amount tagged for a member is usually considered subsidized based on the actual cost of the farm pond. In this model the project ownership could be at a private level. However, at the group level capacity building and monitoring is usually done. The first project that is implemented communally serves as an example for illustrative purposes. After implementation of the communal project, members are usually required to emulate and implement similar ones at the private level or even at the group level. There are certain considerations that are usually made based on a combination or communal and private rainwater harvesting project. These include training the members at the beginning of the project on rainwater harvesting, group constitution, operations and maintenance, agribusiness and marketing. All members are usually expected to coordinate to ensure there is no overlap in the activities.

According to UNEP (2005), in a study covering several countries in Africa, water laws usually do not deal with rainwater harvesting as the direct collection of runoff and storage for later use but rather as a form of disposal of flood water. Despite most of the African countries have had a long tradition of rainwater harvesting criteria, no specific policies have been enacted in order to establish measures to be implemented to advance rainwater harvesting practices. Therefore, lack of policies may impair governance of rainwater harvesting activities and because these activities are usually operated without a governance framework, there are many limitations because while individual village-level projects might succeed in the short term, their long term sustainability may be severely affected in the absence of an appropriate institutional and legal framework at all levels that fosters linkage, exchange and growth.

Even though rainwater harvesting practices are usually part of a national water strategy, it has not been duly prioritized in different countries around the world. This can be shown by limited budget lines allocated from the consolidated funds for rainwater harvesting as opposed to surface water or ground water development investments, which is attributed to attitude and poor understanding of the technology. For instance, a country like Kenya declared rainwater harvesting compulsory for all public primary schools in the year 2006. However, rainwater harvesting policies in Kenya have not been explicit as compared to other parts of the world, for instance India and Sri Lanka where there have been successful rainwater harvesting practices due to explicit policies that have been enacted to govern and properly manage rainwater harvesting techniques.

The social arrangements that produce responsibility in the management of resources are those that create coercion. Such coercion might come in the form of privatization of resources, forcing the exploiter of the resources to bear the cost of negative externalities or government regulation, including fines or imprisonment for nonconforming behaviour. This system according to researchers is referred to as mutual coercion, mutually agreed upon.

Therefore, to improve this situation in rainwater harvesting in Kenya and other countries suffering from water insecurity, there are several steps that can be established such as formulating by-laws and registration of self-help water projects, improving financial facilitation of rainwater harvesting practices, and enhancing management skills through local training. For example, registered groups are usually bound by a country's laws on accountability and this is likely to attract development partners as formal entities.

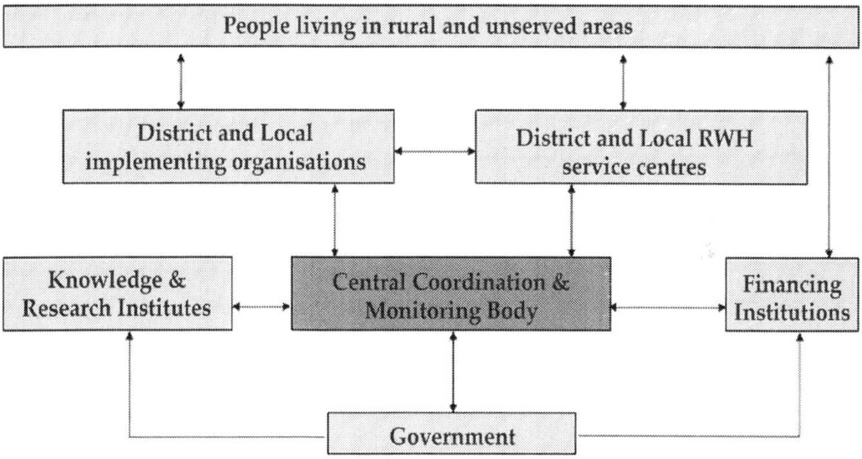

Source: https://www.slideshare.net/ircuser/session-governance-building-an-institutional-rwh-environment-a-schoemaker-rain-foundation

Figure 3.6: An example of a rainwater governance framework

The dynamics and multilevel governance need for institutional linkages across sectors and levels. Vertical and horizontal interplay are important for ensuring institutional fit with social ecological systems. All rainwater harvesting projects for instance would not be sustainable without linkages to the national level actors like the government agencies that provide policy direction and technical guidance, to the banks for credit, to the market for equipment and sale of produce and to other projects for exchange of ideas and solidarity.

Additionally, existence of group by-laws may play a significant role in reducing conflicts or providing avenues for conflict resolution and promoting transparency in the appropriation of benefits. Adoption of rainwater harvesting facilities might not be so popular in most places around world because of insufficient understanding of the technology by many people, hence many who could otherwise benefit from rainwater harvesting may not because of inadequate information.

However, the right to exclusion in the rainwater harvesting situation in a country may not necessarily be vested on individuals but collectively. Exclusion and transfer of rights is a collective decision by the management committee.

TYPES OF RAINWATER HARVESTING AND STORAGE STRUCTURES

According to studies, there are two types of rainwater harvesting structures. These are macro and micro. Some of the macro structures include farm ponds, earth dams, water pans, subsurface, sand dams and roof catchment. Farm ponds and pans are usually considered as diversion, collection and storage structures of rainwater from some external catchment area to a natural or constructed basin structure. Most of these structures may vary based on the size and makeup. In other words, these structures may range from small (manually dug farm ponds) to large community earth and sand dams.

Farm Ponds: Farm ponds are usually more economical than most methods used in rainwater harvesting techniques because they can be used for both farming and domestic purposes and do not require extensive space to install.

Farm ponds are usually practiced mostly by the small-scale farmers though their uses are normally not restricted to micro-irrigation and livestock production. Farm ponds are also used for domestic purposes. Most farm ponds should be constructed by putting in to consideration all the key components that are necessary in reducing water loss through seepage, siltation, evaporation, and improving water management. Some ponds are usually constructed in an integrated rainwater harvesting and storage system. In other words, the system may start from the catchment for harvesting water which is mainly road side or within the farm, the sill trap, screen sill filter, excavated pond and the roof. The technical specifications for the farm pond technology should involve overtime in a bid to make ponds suited to the socio-economic and physical environment.

Source: http://www.journeynlife.com/2014/11/murud-janjira-fort-unconquered-fort-in-india.html

Figure 3.7: A typical pond constructed in arid area for rainwater storage

Sometime NGOs may be involved in the implementation of rainwater harvesting practices such as construction of farm ponds at community level as a way of mitigating various water insecurity issues. One important challenge that may be experienced through involvement of NGOs in the implementation of rainwater harvesting system is the use of imported technology which on one hand may be expensive and on the other unsustainable because of poor compatibility with the social, ecological and cultural environment. This incompatibility may be partly blamed for the failure of many rainwater harvesting projects. Other than the challenge of imported technology, the type of soil in a place can also influence the implementation of farm ponds as a rainwater harvesting structures. For instance, those places with clay soil is possible to line the ponds and retain water for a long time. In this case, it is basically needed to excavate and have a farm pond. However, where the soils are very sandy without the lining, one cannot harvest and retain the water for long. It is therefore crucial for communities to understand the technology so that they can determine when and where to put or not to put the lining in the pond.

Additionally, farm ponds in some places usually require roofing not only to control evaporation and loss of water but also to prevent mosquito breeding and accidents especially for children. Farm ponds are usually durable so long as animals are kept away to prevent piercing and tearing.

Dams and Water Pans: Dams are described as follows "steel reinforced concrete wall which are usually built across a seasonal sandy riverbed." During the rainy season, seasonal rivers usually form and carry soil downstream. The heavy sand accumulates behind the dam, whilst the lighter silt is usually washed downstream over the dam wall. Within one to four rainy seasons most dams are usually filled completely with sand. The water can be abstracted from the sand dam in three main ways,

- Through traditional scoop holes
- Through infiltration gallery either leading to a tank behind the dam or piped through the dam leading to a tap
- Through an infiltration gallery leading to a sealed shallow well in the valley side –topped with a hand pump

Source: https://face2faceafrica.com/article/kenyan-women-innovatively-harvest rainwater-survive-drought

Figure 3.8: A concrete wall dam constructed to assist in the collection and storage of rainwater

Most populations in arid and semi-arid environments such as northern parts of Africa and some parts of Asian countries have been using sub-surface dams and sand dams to harvest rain water.

When implementing rainwater harvesting structures mentioned here (in dams) it is important to consider the issue of water pollution and weigh how to reduce this challenge to sustain rainwater that is of high quality and safe for domestic use. While doing so, it is important to consider installation of a suitable access such as hand pump, so that, instead of people going to scoop water from the sand, they may have a well-equipped hand pump for easier access and this would significantly help in reducing pollution. In other words, it is apparent that once a dam is installed, it is significant that a hand pump is erected to assist in drawing water from the dam in a more hygienic and safe manner. In addition, an animal trough should be constructed on the side so that livestock can access water from this point instead of going directly to the dam. This helps in preventing contamination from animals. Generally, equipping a dam this way improves the quality of water making it safer for domestic use. This is also emphasized in different literatures because one of the challenges facing rainwater harvesting projects, especially dams, is contamination and subsequent quality compromise.

Source: https://dannyfarm.wordpress.com/tag/rain-water-harvesting/

Figure 3.9: The body of water behind the man is an example of a water pan for rainwater harvesting and storage

Water pans are described to be important structures that requires a tractor or a bull dozer to impound the soil without much compaction hence, creating a reservoir. Water pans usually do not have a regular shape; however, they are less intensive in terms of effort in excavation. One situation that should be considered when constructing water pans is that water will be available to be used by the community and most cases it will not go around the season, because after three months, the pan will most likely to have dried up. Compared to other structures used in harvesting and storing rainwater, pans are normally smaller than dams and much shallow than ponds. One of the challenges associated with dams and pans is that they are not considered as sources for potable water. It has also been indicated that this is usually the case because of the way such projects were implemented. For instance, many contractors usually cut down the expenses thus ending up installing the project halfway. Usually, dams and pans become an issue of just excavation whereas there is much more to do after excavation to meet the required portability standards.

When constructing a water pan to provide clean water to the community, the contractor is supposed to set-up a filtration system on the upper side of the whirl, that is, the wall and the filtration system is composed of ground material of rocks, small and big rocks and then sand. They are put in layers such that the water that is going through it is purified. However, this system may not be very effective if the water is very turbid. In order to ensure that the community can get clean water from this turbid water beside the filtration system installed, another important system is attached to it, commonly referred to as slow sand filter which allows water to pass through another filtration stage. This system is composed of sand, boulders, ballast ground such that water will go through that and the volume is big. By the time it comes out on the other side, it is usually cleaner than it was before.

United Nations Environmental Protection (UNEP, 2005) suggest that the success and sustainability of rainwater harvesting interventions such as dams is influenced by identification of suitable sites. Apart from a site being technically suitable for a certain intervention one must also take into account the social economic and technical requirements. A site could be suitable for a dam but far away from a settlement. Thus, there is a need to integrate the technical requirements with the socio-economic conditions of the beneficiaries because it is the people who would manage and utilize these structures. Integration of the technical requirements with the socio-economic condition of the beneficiaries would therefore indicate optimal site for such interventions.

Roof Top Catchment: Roof top catchment is one of the most common of all the rainwater harvesting techniques that are practiced in different parts of the world. This includes structures in both urban and rural areas. Roof catchment water is often used for domestic purposes. In conjunction with roof top rainwater harvesting methods, various types of containers are usually used to store water from the roof including clay jars, drums tanks and under/ground reservoirs.

Source: http://lms.sjp.ac.lk/fas/wordpress/?p=329

Figure 3.10: Roof top rainwater catchment

Most rainwater storage tanks are usually made up of corrugated galvanized iron sheets in most parts of the world. However, most of these tanks are usually abandoned due to leakages resulting from corrosion. Some places also use Ferro-cement tanks in storing rainwater. However, many problems are usually experienced with the use of Ferro-cement tanks because many tanks are constructed poorly using poor sand and inadequate curing, which results in porous walls that give way to seepage. It has also been proven that Ferro-cement tanks are the most popular material for construction due to its affordability and quality. Most of the Ferro-cement tanks found in most homesteads are usually unutilized due to poor workmanship.

Some of the limiting factors to the use of roof catchment as a mode of rainwater harvesting method are the size of the catchment, the capacity of the harvesting container and awareness on the other technicalities of the system. In other words, most average people have iron roofed houses but the house could be small and the tank is small, thus limiting the optimal harvesting and storage of rainwater. In any case, the idea is to increase the size of rainwater harvesting catchment area (roof-top) and the size of storage rainwater tanks.

According to UNEP (2005), a well-constructed roof top catchment is composed of a fowl flash and a well-insulated tank that will provide safe water for domestic use. A fowl flash ensures that no dirt particles enters the tank. If the tank is completely closed and shielded from the sun, then the microbes are not able to grow and the water remains safe.

Considering the cost incurred to construct a rainwater tank, it is therefore significant to have other alternatives such as building a dam which to some extent is much cheaper compared to a tank. When constructing the dam, it is advisable to carry out a calculation to find out how many tanks or dams will be needed to sustain water supplies throughout the year. Many people do not estimate and just construct a tank and then harvest from the roof. You need to know the surface area of your catchment. You need to know the demand per day and then relate to that storage and then, gather data as to the number of times it will rain. This will allow the designer to estimate if the facility can support the community or a household's water consumption over a period.

Rock Catchment: A rock catchment can be described as a gathering system with a concrete wall built on top of a rock. This normally happen where there is a big surface area of rock. However, this water does not usually last long. It is likely to be exhausted within two to three months after the rains due to evaporation. Hence, collected water should be channelled through a sand and gravel filter to a tank.

Source: http://ke.geoview.info/malilu_rock_catchment,87209492p

Figure 3.11: Rock catchment for rainwater harvesting and storage

For this system of rainwater harvesting, there is the need to carry out maintenance of the rock catchment. This included erection of a fence so that animals such as goats and cows don't climb the rock and cause contamination. During dry season, dust may gather on top of the catchment. Therefore, the management group or caretaker of the project should be trained to be on the lookout so that the first rain is not allowed into the tank to clean the surface and get rid of the dusty water.

Though rock catchment is limited in site availability, rock catchment can collect significant water within small showers just like roof catchment but it is also limited in use (to domestic use) as compared to open reservoirs like dams unless the water is distributed/channelled to points where it can be used for irrigation and livestock. Compared to a dam, the water quantity collected through rock catchment is usually smaller hence, it cannot in most cases, suffice for farming. It is thus, mostly used for domestic purposes.

In-situ Rainwater Harvesting: In-situ rainwater harvesting technologies that refers to the trapping of rain water where it falls on the ground for maximum soil filtration. This is usually done through techniques like terracing, pits and ridges. In most cases, two or three in-situ technologies can be used at the same time, for example, terracing plus zai pits or contour ridges plus zai pits to ensure maximum filtration.

For a long time, in-situ rainwater harvesting has been considered crucial for conservation agriculture hence, it should be relegated to the Ministry of Agriculture. According to experts, the process involved is collecting and storing rainwater within the soil and therefore, should be considered as an important method of rainwater harvesting.

CONCLUSION AND RECOMMENDATIONS

Rainwater harvesting has for a long time been practiced widely and is one of the major sources of water supply in most parts of the world, particularly in arid and semi-arid areas. Studies have shown that different communities in the rural arid and semi-arid areas have been using rainwater since time immemorial. The technologies used by different people in rainwater harvesting may vary from one user to another. Some of these variations included cost, intended use of water and amount of rain received. Additionally, rainwater harvesting in most places have however, been being operated below expectations and perceived potential.

The main reason for the low performance of rainwater harvesting is the absence of effective policies and legislations in a country and the consequential lack of institutional support for innovation, implementation and local management. The lack of rainwater harvesting policy is to a large extent responsible for the meager resource allocation by the government and poor coordination in rainwater harvesting sector. Different studies have also shown that policy inadequacy affect rainwater harvesting progress. However, the policy environment around rainwater harvesting is likely to become more favorable if effective water acts are to be enacted, which specifically should greatly focus on rainwater harvesting development. Additionally, responsible ministries such as the Ministry of Agriculture and Ministries of Water and Irrigation, together with other responsible agencies should be committed towards enhancing various rainwater harvesting processes. There should be a comprehensive policy covering all aspects of rainwater harvesting including research and development, governance and funding.

Capacity building for rainwater harvesting is needed at all levels, especially in arid and semi-arid areas. However, the capacity of management groups is specifically important for sustainable rainwater harvesting projects. It has been stated regularly that without knowledge and awareness, people are not able to demand or even if they know a little bit, the inadequacy of technical know-how and important information of rainwater harvesting and storage can mostly make them inefficient managers and monitors. Promoting effective rainwater harvesting efforts therefore, needs to start from effective local institutions and to establish effective local institutions, sound mobilization and training is crucial to promote social cohesion that is instrumental in the strength of local institutions. Cohesive local institutions are mainly associated with successful rainwater harvesting. Studies have shown the importance of strong institutions, capacity and cohesion in ensuring sustainability. Even though there may be no project group that can be functioning perfectly, it is apparent that some of the groups may be functioning more effectively than others thus, promoting sustainable rainwater harvesting projects. The characteristics common in these groups may include; existence of enforceable rules, clear benefit sharing strategy, sense of ownership, addition of value chain, capacity, monitoring system, and social capital cultivation. These characteristics may be contrasted with those common to non-successful groups which are usually characterized by conflicts, weak teamwork, degraded projects, corruption, domination by cliques and unclear ownership rights.

Therefore, poor linkage and coordination among rainwater harvesting institutions may result in duplication of efforts and lack of scaling up of rainwater harvesting activities in the country.

There is also the need for both vertical and horizontal linkages across levels including local, national and international. In other words, coordination of resources (skills, finances, learnt lessons) at all levels should be enhanced through creation of a national rainwater network which would generate more linkages and synergy. Literatures have also supported networking and creation of vertical and horizontal interplay to address poor communication and duplication challenges. The national network would facilitate information sharing among various actors (experts, donors, consultants, researchers, industrialists, policy makers), capacity building, oversee certification and standardization, address duplication, quality and capacity issues in rainwater harvesting processes. To complement this, a multilayer implementation structure that would link local rainwater harvesting projects groups from the sub-locations through locations to the national network should be promoted.

There are several principles that should be incorporated in any rainwater harvesting project. These principles are discussed below in detail.

The first principle entails embracing an application of a combination model in rainwater harvesting water supply projects. Inadequacies of water from the rainwater harvesting projects have been the biggest obstacle in most places around the world. This is mainly attributed partly to the capacity of containers/reservoirs, siltation or disconnect between quantity and demand. It is therefore recommended that a combination model should be adopted, especially where cost is an issue. This for instance, ensure that a bigger capacity reservoir (dam, pan, and pond) is utilized for production while a roof catchment is installed for domestic purposes.

The second principle entails an application of standards. All the promoters involved in rainwater harvesting projects are recommended to adopt standards in their practices, such as, the WHO standards for water supply in schools and SPHERE project's minimum standards for water supply and sanitation, all through the project cycle to ensure effectiveness of rainwater harvesting interventions. These are internal standards that specify for instance amount of water required per person per day. One of the important factors that contribute to failure of rainwater harvesting project is the installation of projects without adherence to recommended standards. Standards for rainwater harvesting projects should start with baseline surveys, feasibility studies which provided data necessary to support adoption of standards on quality quantity, as well as time and distance as far as rainwater is concerned. An initial feasibility study is important because it may help to indicate the viability of an intended project based on anticipated quantity from the mapped catchment whereas baseline survey data would provide details on the status of the intended beneficiaries and the water demand amounts. This information would help ensure that the beneficiaries are able to access the required/recommended amounts of water per day, and at the recommended maximum distance and time taken to the water point.

The other principle that should be recommended is on the regulation of rainwater harvesting infrastructure and designs. For rainwater to be regarded as potable, there is the need for regulation of the way in which the infrastructure is installed and water-handled. For instance, water from dams, and pans was regarded as non-potable. Most often this is usually the case because the rainwater harvesting and storage structures are not installed to completion. Many dams are usually abandoned after excavation and regarded as complete. However, for the dam to provide clean water, more than excavation is required. For instance, fencing, sinking a well or installing a filtration chamber, separate water points for domestic, livestock and production, a water pump and sanitation facilities in the surrounding, distribution and water treatment facilities must be considered. Roof catchment could also be improved through installation of a foul flush system to ensure that only clean water gets into the tank. Treatment of rainwater whether at a central place or at the household would improve the water quality. This in the essence, would make the structures more expensive and that is why many contractors and sponsors usually take shortcuts. Therefore, with regulation by the government, it would be possible to immensely improve the quality of rainwater. However, these guidelines are rarely adhered to intentionally or unknowingly. Consequently, existing guidelines ought to be disseminated and those that are missing be undertaken.

It is important to focus on the utilization of Local materials in enhancing rainwater harvesting process. One of the challenges that can be noticed in rainwater harvesting projects is the inability to take care of operations and maintenance which may render projects unsustainable. This is usually caused by factors, such as, utilization of imported materials which made repairs difficult due to inaccessible spare parts and lack of involvement of local people in project planning resulting in the failure to transfer knowledge and capacity. Therefore, all sponsors of rainwater harvesting projects should be recommended to use locally available materials, or customize imported models using local materials and build the capacity of the beneficiaries to maintain the projects.

Other recommendation that can be given towards enhancing rainwater harvesting projects include encouraging sensitization and community responsibility, proper networking and coordination in rainwater harvesting technologies, encouraging academic and research institutions to develop curriculum on rainwater harvesting, improving financial resources in rainwater harvesting projects, and enhancing institutional design in order to support rainwater harvesting and storage in arid and semi-arid areas.

4 GOVERNANCE IN WATER SUPPLY SYSTEMS IN THE INFORMAL SETTLEMENTS

INTRODUCTION

Governance affects the disparity in access to water supply in a country. Majority of people, especially in informal settlements are not able to have a well-structured network of water supply system due to failure of governance. This critical problem of governance usually plays a significant role in perpetuating contemporary crises in water supply in a place. Distribution of water supply network is highly correlated with household income. For instance, it is a common sight where large water mains run through a poor community yet the adjacent homes do not have outlets to access the water. This brings about critical concern for equity and the right to access basic water resources in the informal settlements. The poor are particularly highly vulnerable since governments and local authorities are faced with financial and institutional challenges. It is therefore important to evaluate various factors that influence access to water such as water quantity, quality, costs, distance, and time.

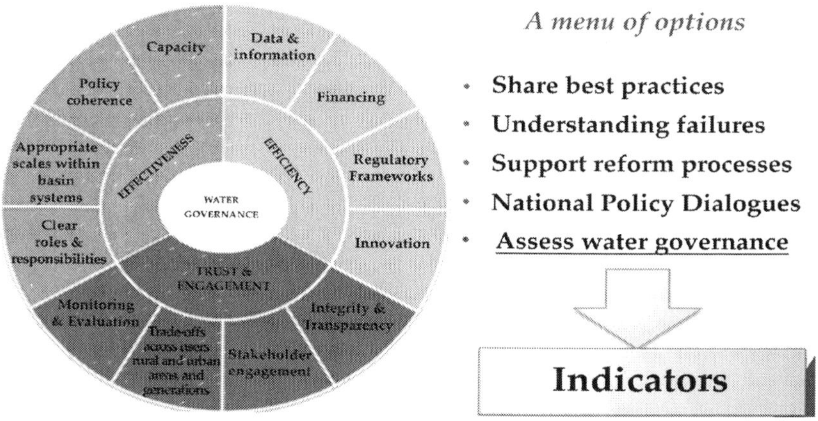

Source: https://www.slideshare.net/OECD-GOV/stakeholder-consultation-on-water-governance-indicators-oecd-secretariat

Figure 4.1: Principles of water governance

Studies have shown an increase in the rate of urbanization in most developing countries, especially in African and some parts of Asian countries. This signifies that informal settlements will continue to grow at an exceptional rate and thus increasing the demand for utility services like water supply systems. Amidst these trends, the inability of water service providers to provide adequate and affordable water has created ideal conditions for illegal water vendors to thrive. The resulting situation is that households have limited sources of water with limited quantities and generally unacceptable quality levels. Moreover, households pay significantly higher costs for water than other households in well-to-do areas. These are indicators of ineffective water governance and have a direct impact on how households access water.

It is therefore important to understand the sources of water in informal settlement, the actors involved in the provision of water and the challenges faced by both households and actors. Through an evaluation of these challenges, an effective water governance framework can be formulated and as a result, might lead to improved development outcomes in other water-related sectors like health and education.

MAIN SOURCES OF WATER IN INFORMAL SETTLEMENT

Studies have revealed that more than75% of people living in the informal settlements obtain their water from water vendors. It has also been revealed that Water vendors are usually the preferred source since the water quality is acceptable and the cost of water relatively is affordable to most households. About 15% of households obtain water from public taps. These are taps set up by the water utility companies at strategic points from the main water network connection. The dependency on this source is usually low since most of the time such taps are vandalized as well as frequently diverted to illegal connections. About 7% of households depend on piped water directly into their households from the utility companies. This source is only available to areas around the periphery and close to the main transport corridors where it is relatively difficult to divert water. Further researches have revealed that this source is usually unreliable and experienced rampant water shortages. Another 3% of households usually access their water from other sources such as tankers and boreholes. Boreholes are the least preferred since there is a general impression among residents that the ground water in informal settlements is highly polluted. Water tankers on the other hand, cannot access most parts of the informal settlement due to the poor state of access paths and uneven terrain, which is characteristic of informal settlements.

Source: http://www.eawag.ch/en/department/ess/projekte/business-innovations-for-making-water-kiosks-more-successful/

Figure 4.2: Residents in the informal settlements fetching water from the water kiosk

FACTORS INFLUENCING WATER SUPPLY IN INFORMAL SETTLEMENTS

Water Demand: According to World Health Organization (WHO), it has been revealed that an average household in the informal settlements is made up of 5 persons.

Therefore, the demand for water in households located in informal settlement is relatively high. Research findings show that majority of people in these areas usually consume between 60 and 200 litres of water daily for their domestic uses. It is also worth noting that the most residents in the informal settlements always diversify their sources of water depending on the different uses. For example, water intended for laundry is usually obtained from tapped systems, while water for cooking and drinking is individually obtained from the water vendors and this is usually treated first before use. Figure 4.3 shows the average amount of water used per day by the residents in the informal settlements.

Statistically, the average water usage in a place is usually computed based on litres per capita per day. This can then be compared to the average minimum for basic health protection of less than 20 liters per capita per day, of which 7.5 litres is required for consumption as recommended by the World Health Organization. These levels of access can also be interpreted in terms of household water security.

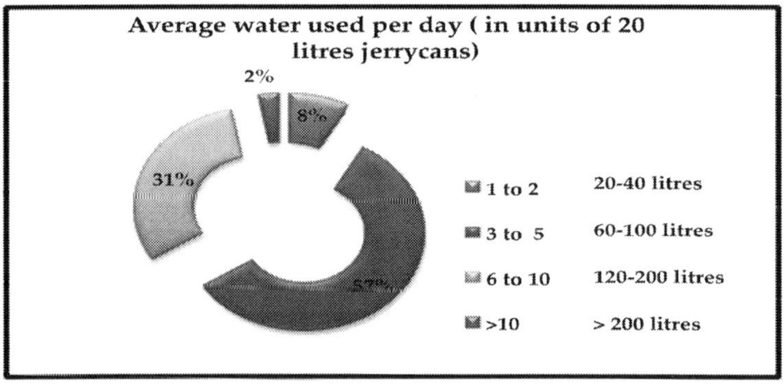

Source: http://www.scielo.org.za/scielo.php?script=sci_arttext&pid –S1816-79502010000500009

Figure 4.3: Average amount of water used per day per household in the informal settlements

The average water use of 20 litres per capita per day in the informal settlements barely meets the service level description of basic access. The collected water quantity is classified as low where other domestic chores such as laundry and bathing may occur at water sources with additional volumes of water. Similarly, the quality of water in this category cannot be assured. It is also important to note that these figures are indicative and do not consider the age, gender or occupation of persons in a household. Different categories of persons have varying water demands.

Institutional Issues: The deteriorating quality of services and the poor governance of water service delivery can to a large extent be blamed on institutional problems such as inadequate political will and commitment, and the lack of a long-term, comprehensive, development vision, all of which discourage investment in water sector.

Political commitment to ensuring potable water delivery should be demonstrated through actions to improve coverage and delivery, as well as actions to ensure the efficient and equitable provision of services. This would include, for example, the creation of an enabling environment for improved service delivery, such as by allocating funds for research and development, and through the adoption and enforcement of a legal framework with a minimum standard for service delivery. This should logically begin with a definition of a minimum standard of potable water service. For instance, in the informal settlements, the minimum standard in water delivery services encompasses factors such as quantity, pressure, distance, reliability, and education on wise use. It includes:

- A minimum quantity of potable water of 25 litres per person per day or 6 kilo-litres per household per month
- At a minimum flow rate of not less than 10 litres per minute
- A maximum of 200 metres of a household; and
- Effectiveness such that no consumer is without a supply for more than seven full days in any year

Unfortunately, political intervention has all too often been transformed into interference for political gains. In most parts of the world, coverage and improved water service delivery have been politicized and used as an electoral tool to gain popular votes. For instance, in Africa it has been observed that the recruitment and advancement of senior administrative and management staff in the water sector is politically driven rather than based on professional aptitudes. In some countries, politicians have meddled in water pricing and have advocated for free supply or very low tariffs to the detriment of service quality to gain popular support. Local politicians have also contributed to the mismanagement of this sector by the over-staffing of the lower levels with unskilled workers.

Thus, political meddling in the water delivery sector contributes to the lack of skilled staff and results in low operational efficiency due to, for example, a high ratio of staff per water connection or consumer served. The unethical practices and malpractices (such as assisting in illegal water connections, manipulation of meters, falsifying meter readings, and asking for bribes in return for services) attributed to those politically appointed unskilled employees affect the public image of the utility and further compounds its revenue problems. Political meddling also fosters a culture where lack of accountability and transparency is acceptable and can lead to corruption, absence of the rule of law and other unethical practices, which in turn, can aggravate the already poor performance of water sector. The impact on quality of services, willingness to pay, and cost recovery by the utilities is great, thereby contributing to the problem of lack of funds required to improve such services.

Another aspect of institution issues involves lack of long-term integrated development planning. The capacity of water delivery systems can be stressed when total water demand exceeds the supply capacity. This can be due to population growth, urbanization, or a decrease in source water productivity. Population growth and urbanization generally occur gradually, or if spontaneous and unprecedented, are due to a development initiative, such as the creation of a new industry or school, or to the dislocation of a population. Logically, if a development initiative is conceived and executed within a comprehensive development agenda rather than as a quick-fix political solution to address a long-neglected development imbalance, water-related problems will be easier to address by rectifying existing management techniques. The problems on population growth and the corresponding increase in demand for social services in general, will be more targeted-water demand will be properly predicted and the resources will be correspondingly allocated.

The lack of long-term integrated planning vision can also be demonstrated in situations where poor quality services are the result of inadequate and aging infrastructure due to the lack of revenue and/

or population growth. A long-term integrated planning vision would have accounted for proactive repairs, maintenance, expansion, and replacement through amortization. The current deteriorating state of water infrastructure and the necessity for major repairs and replacements suggest that past development and planning has failed to consider the life-cycle or resource constraints or that there was a chronic lack of follow-through on promises, follow-up on projects or routine maintenance work. These infrastructure and institutional challenges have not gone unnoticed. Attempts have been made by different countries to improve the above-mentioned problems. State-level reforms have been enacted with a goal to improving the state of infrastructure, revenue, quality of services, as well as management and governance practices.

Distance to Water Source: Based on the proximity of water source, studies show that about 55% of people in the informal settlements usually access water within a distance of 500 meters while 40% of them usually cover between 500 to 750 meters to their nearest source of water, while the other 5% cover between 750 meters to 1 kilometre to access water. This signifies that over half of the households in the informal settlements usually have to access their water beyond a radius of 500 meters. This is significantly long distance when other factors such as insecurity and time constraints are considered. The service level description as defined by UNEP (2005) categorizes the distance between 100 meters and 1000 meters distance as a zone of basic access to water. In this service level, not all water requirements may be met.

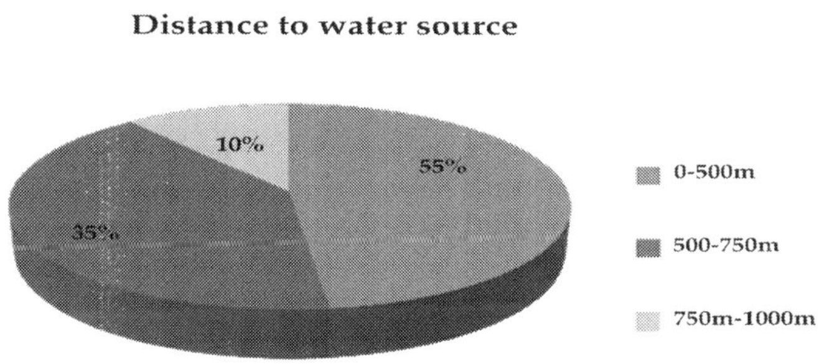

Source: https://www.researchgate.net/figure/313216189_fig2_Figure-2-Distance-of-fresh-water-source-for-irrigation-from-farmer%27s-crop-field

Figure 4.4: Distance to the nearest water source in the informal settlements

Time Taken to Access Water: The time taken to reach the nearest water source is highly significant in informal settlements as this may not always depend on the distance. There are other factors that determine the time taken to reach the water source. These include the accessibility of the paths, since most of them are usually narrow, and the volume of people queuing for water at a given source. The time of the day is also significant since most residents draw water during early morning hours and late in the evenings. Security of the residents is also a determinant factor since some routes may be shorter in distance but riskier especially to women and girls. Research findings have revealed that 30% of the residents in the informal settlements usually take less than 5 minutes to reach the water source. Most of these residents however usually have individual piped connections (legal or illegal) or lived close the water points. About 35% percent take between 5 to 10 minutes to reach the water source, 20% take 10 to 20 minutes while about 10% take between 20 to 30 minutes. Finally, only 5% usually take more than 30 minutes and these are found to be negligible incidences. From the analysis of service level description, the range between 0 to 5 minutes is considered acceptable, but this also leads to an increase in the quantity of water consumed.

Infrastructure Issues: Water delivery infrastructure in the informal settlements, especially in developing countries is as varied as the types of supply paradigm. They include open water sources with no infrastructure requirement, wells and boreholes with simple metal casting lids, rope and bucket wells, improved wells with pumps and single or multiple taps, and urban piped water delivery distribution systems with several types of infrastructure requirements. Potable, piped-water delivery system infrastructure in urban areas typically includes intake structures, treatment facilities, transmission mains, storage tanks, pumps, valves, and distribution networks. Rapid population growth of 2.5 per cent and rampant urbanization estimated at 4.4 per cent have negatively affected utilities that rely on piped water delivery because most of the growth occurs in slums and unplanned areas considered to be risky for the expansion of networks or where return on investment is unattractive.

The state of water delivery infrastructure in the informal settlements varies depending on the supply technology, ownership, management, and governance approaches employed in each case. For instance, the specific technology and its complexity depend to a large extent on the investment and management capacity of the individual service provider. There are several elements that should be considered when discussing issues of water supply infrastructures. For instance, there are the issues on inadequate and aging infrastructures.

Inadequate infrastructure can be broadly defined as a situation where the existing infrastructure is unable to meet consumer demand with respect to both quantity and quality of services. Low coverage (a serious problem in the informal settlements) is a typical indicator of inadequate infrastructure. Rapid population growth can generate an increase in water demand that can stress the capacity of existing infrastructure.

The health, social, and economic impacts of inadequate infrastructure have been well documented. Inadequate infrastructure can result in intermittent or irregular supply, poor water quality, and low or insufficient levels of usage. It also promotes the use of unsafe drinking water sources (as an alternative) and aggravates poverty in several ways. For example, in such cases where illness results from the use of unsafe water or from the inadequacy of supply, the loss of productive work days and the money spent on medication is significant. Furthermore, the cost of buying water from vendors is usually very high, and the long time spent in collecting water results in the forfeiture of important social and economic activities, such as care of children by mothers and quality of family life and, for children, results in tardiness at school and reduced productivity due to fatigue from their water fetching duties.

Aging is a natural process characterized by the decline in the quality of materials due to corrosion or rust. Infrastructure in bad condition, such as when materials have exceeded their life span, is of concern in potable water delivery because of increased commercial losses for utilities and high rates of leakages. In Ethiopia, for instance, about 70 per cent of commercial losses are attributed to aging infrastructure.

Source: http://efc.web.unc.edu/2016/03/24/aging-infrastructure/

Figure 4.5: Aging water supply infrastructures

Aging infrastructure is also of concern because of the high cost of maintenance and the risk of frequent breakdowns, which together can worsen the poor revenue returns for water utilities and thus reduce the quality of their services. Aging infrastructure can also contribute to public health risks from bacterial re-contamination of treated water and the chemical toxicity of heavy metals that may leach from old pipes. Moreover, it can contribute to the rapid depletion and ineffective action of the residual chlorine in the distribution networks because the breakdown in the building materials provides favourable conditions for bacterial growth. In addition, the extra sediments that are deposited in the water also screen bacteria from the action of disinfectants.

The situation of inadequate and aging water delivery infrastructures cannot be considered in isolation, but points to bigger challenges related to institutional problems in water management and governance practices.

Cost of Water: The cost of water is a major factor in the determination of various domestic activities to be undertaken. This might also take up a significant proportion of the household income. Since this is a daily expense and not a monthly expenditure, it constantly triggers the residents' consciousness on the limit of how much to spend in each day. For instance, in a country like Kenya, it has been revealed that about 45% of the residents in the informal settlements usually pay between

10 and 20 Kenya shillings per unit of a 20-litre jerry can. This is the average range charged per unit. Additionally, it has been showed that the cost of water usually depends on the source of water and the prevailing weather season. Borehole water is relatively cheaper compared to the treated water from the water vendors. In other words, water supplied by vendors to households is usually charged at a higher cost (between 10 to 20 shillings) than water individually fetched by households from other sources. It is also worth noting that some water vendors only sell water to other smaller vendors who in turn sell the water to households at an unreasonably high fee. These are mainly what households refer to as cartels. In the event of water scarcity, or dry weather periods, water prices escalated arbitrarily and water vendors and cartels usually cash in on the households' misery.

The Quality of Services: The quality of water supply service can be a composite of several factors: reliability of supply, water quality, accessibility, water pressure and time spent in collecting water, as well as customer service. These challenges that characterize the quality of service for water utilities in a place are discussed in the following sub-sections.

The first issue on the quality of water service entails intermittent and unreliable water delivery. According to World Health Organisation (2010), intermittent water delivery refers to a method of water scheduling, which consists of rotating the supply of water to the different areas of a distribution network (regulating days and number of hours of flow), that has been adequately communicated to users. This distribution method is used as a strategy for equitable water delivery, especially in water-scarce regions and/or during periods of water scarcity. Although it can stress users by reducing their water use and consequently the number of basic needs that can be met, if adequately communicated and respected, intermittent water delivery can be less burdensome to users than unreliable supply (as defined below), because it allows them to develop adequate coping mechanisms, such as planned water storage and the prioritization of water use.

One of the coping strategies that can be used is using proper water storage employed when the provision of potable water is intermittent or unreliable. This is because most users do not normally store water in situations of continuous flow. This practice has been documented as common and widely used by users in several developing countries and in Sub Saharan Africa. Studies have revealed that about 20 per cent of the population, especially those in high- density, low-income areas, usually receive water supply services for as little as one to five hours per day, while another 40 per cent of users receives services for less than 12 hours per day.

Unreliable water delivery refers to a situation where the water provision to a particular area begins and then is interrupted without following a defined schedule known to users. The service is interrupted either due to a breakdown in the water delivery system or due to poor communication of the water scheduling. Unreliable water delivery can result in significant negative consequences for individuals and for society at large. Firstly, it can cause water -related diseases that result from the poor hygienic conditions of containers used to store drinking water or from the consumption of poor-quality water from unsafe alternative sources. Secondly, unreliable water delivery can promote water wastage, and can result in inflated bills and the consequent lack of willingness to pay such bills. Stored water, for example, is usually inefficiently used or discarded for fresher water when flow resumes. It also triggers the practice of leaving taps open during periods of flow interruption which has been known to cause significant flood damage both in homes and in public spaces when flow is re-established during the night or when nobody is available to close the tap. In addition to such damage, it may contribute to inflated water bills, which affect the users' incomes as well as the providers' cost recovery due to users' unwillingness to pay such bills, and the wastage of an already scarce resource further aggravates the water shortage and necessitates further rationing.

The other issue entails poor water quality. According to WHO (2010), to protect the health of users and to ensure public welfare, raw water should be treated to eliminate or reduce the concentration of any substance that poses a health risk. Of concern, is the presence of faecal matter that can lead to severe epidemics and widespread death, causing huge social-economic and political upheaval. The presence of E-coli in water, for example, is an indication of faecal contamination, and must be completely absent from potable water. The lack of sufficient residual disinfectant in a water distribution network poses a high risk to users in the event of E-coli contamination of drinking water delivery, such as what has occurred in Peru, South Africa, and in Walkerton, Canada. In addition, an adequate quantity of residual disinfectant in a given distribution network is necessary to preserve the quality of treated water in the event of recontamination during its transmission and distribution. Also of concern are the presence of naturally-occurring substances, such as heavy metals, cyanides, arsenic, or pollutants from human activities such as from the manufacture of pesticides and other organic substances, that have chronic consequences on human health.

According to WHO & UNICEF (2000), the provision of unsafe drinking water is common in most of the developing countries due to no or inadequate treatment of raw water. Although this is generally attributed to the lack of funds necessary to purchase disinfectant chemicals, negligence in monitoring the level of residual disinfectant in each network, as well as the widespread belief that groundwater is free of contaminants, are also to blame.

The other issue involves accessibility to water source point. For instance, a large majority of users in the informal settlements usually face the enormous burden of walking to collect water from piped networks. According to UNICEF (2010), in developing countries, only 35 per cent and 5 per cent of the urban and rural populations respectively, have water piped directly into their houses. However, the actual percentage is lower if unreliability of supply is considered. These statistics indicate that many households do not have improved water delivery sources on site, and therefore their water delivery needs are met via lower service levels, such as public standpipes or alternative sources such as springs and streams, that are located some tens of metres to several kilometres away from point-of-use.

Long walking distances to water delivery sources affect the quantity of water consumed and consequently the levels of personal hygiene and sanitation, and the disturbing reality is that the household water collection chore falls mostly to teenagers, who should make several trips per day to collect water in containers of average volumes ranging from 20 to 50 litres. These youths spend a significant number of hours fetching water, with the forfeited alternatives being academics and leisure activities. This pattern is especially true for female youths (WHO/UNICEF, 2006).

Another issue is the inadequate water pressure and long waiting times at the tap. The discharge rate of water flowing from a tap is an important indicator in measuring the burden of water collection and utility performance. Inadequate water pressure means a longer time to fill each container, generating long queues or overcrowded taps, and thus results in longer waiting times at the point of collection or at public standpipes. A maximum of 3 minutes has been recommended as the time necessary to fill a 20 litre container. Waiting times of up to 4 hours at public standpipes have been reported during peak hours, early hours of the day, and late evenings due to overcrowded taps in most parts of developing countries such as Africa. Long waiting times at improved sources is especially problematic where unsafe alternative sources are readily available since this can trigger the use of such unsafe water sources.

Source: http://www.humanosphere.org/basics/2013/01/the-problem-of-cheap-water/

Figure 4.6: Long queuing line for water at the source point

Lastly, there is the issue to do with lack of customer service by utility companies. Customers deserve the opportunity to provide feedback on the quality of services received, as well as to contribute ideas to and inform management decisions. They also deserve timely repairs and maintenance, to be adequately informed of the scheduling of services, as well as of any interruptions, and should experience ease in making payments for such services. More than 90 per cent of water utility companies in developing countries do not have a customer service culture.

Perhaps due to the monopoly in water supply service provision, consumer service has received the least attention from water delivery authorities and it has continued to deteriorate in the face of increasing demand for improved services. In most cases, the procedures and mechanisms for complaints are non-existent, and, where available, they are complex and highly inequitable. Further issues of concern include faulty meters, single payment points which generate long queues, incorrect meter readings, non-credited payments, long connection and reconnection times, and poor attention to leakages unless the customer provides some form of (cash) incentives to employees of the water utility. For instance, in Benin, Cameroon, Lesotho, and Malawi, it takes more than 2 months to get connected, whereas it takes less than 20 days in Ghana, Ethiopia, and Namibia.

Poor quality of water supply services affects both the users and the utility companies. When users are not satisfied with the quality of services received their willingness to pay decreases. Reluctance to pay may generate high rates of non-payment of bills and consequently diminishes the revenue base of the utility. This can generate a vicious cycle of poor service, non-payment and a correspondingly weak revenue base. The inferior quality of services can be linked to infrastructure problems, as well as to poor management and governance practices in the water sector.

WATER GOVERNANCE IN THE INFORMAL SETTLEMENTS

Water Actors in the Informal Settlements

There are usually different actors involved in the provision and distribution of water in the informal settlements. Most of these actors can be classified as either formal or informal actors. Formal actors are mainly those that are licensed and therefore legally recognized to operate in the area. Informal actors involved those who are not licensed but carried out the business of selling water to households in the area. The following is a breakdown of various actors involved in the provision of water in the informal settlements and their main characteristic.

- *Municipal councils* are classified as formal actors. They mainly supply water through piped networks. They usually have a minimal spatial coverage. They also experience frequent interruption of water supply. The quality of water supplied is usually acceptable but further treatment is required due to contamination from illegal connections.
- *Water kiosks* are mainly classified as informal water actors. This actor supplies water through municipal piped networks to a common water point. Once the water is supplied, the water kiosks then resell the water to households and vendors at relatively affordable costs. Most water kiosks usually have water meters and are required to pay water bills to the municipals. The quality of water is usually acceptable but not fit for human consumption without further treatment.
- *Cartels* are informal water actor. Most often are masqueraded as water kiosks but are usually unlicensed. They obtain their water by making illegal connections from main water supply network. They also have no standard water charges (usually fluctuates with demand and supply).
- *Water vendors* are classified as informal water actors. They usually obtain water from water kiosks and cartels. Most often they use hand carts and wheelbarrows to deliver water to households. The quality of supplied water is not acceptable due to containers used for water delivery.
- *NGOs and CBOs* are also classified as formal water actors. They usually provide water to all households in the informal settlement areas. They also have huge capacities for water storage. Some usually incorporate other income generating activities alongside water provision e.g. youth projects, sanitation blocks etc.
- *Self-help groups (SHG)* are another group of formal water actors. They usually provide water to member households. They also have limited capacities for water storage compared to the NGOs and CBOs. They also incorporate other income generating activities alongside water provision e.g. youth projects, sanitation blocks etc.

Management of water sources in informal settlements is critical as this reveals the inherent challenges of governance such as the level of service provision, nature and level of participation in decision making, and other factors like marginalization of specific gender especially women.

According to World Health Organization (2010), about 45% of the water supply systems in the informal settlements are usually managed by water vendors, while cartels are responsible of about 25% of the water supply activities. In essence, about 70% of supply systems are owned and managed by water vendors and the cartels while 20% of the water points are usually under the management of self-help groups. The municipal companies are only responsible of only 10% of the water outlets in the informal settlement. This is in contrast to the ideal role of the water company since it is legally tasked to provide adequate water to all citizens within an area.

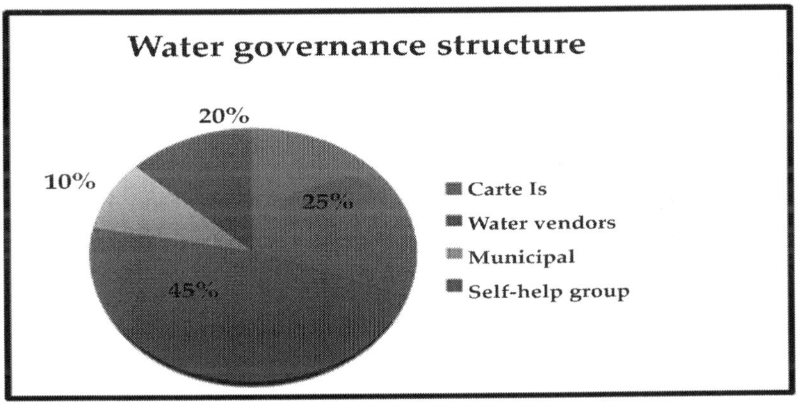

Source: https://www.researchgate.net/figure/228465388_fig3_Figure-3-Primary-source-of-water-in-Kisumu-city

Figure 4.7: Water governance structure in the informal settlements

Most residents in the informal settlement usually prefer Community Based Organizations (CBO's) and Non-Governmental Organizations (NGO's) to control the water points in the area because of different reasons. For instance, these organizations are best placed to understand the water needs of households in addition to being non-partisan in provision of water. Most self-help groups are usually faulted for being discriminatory and only advocating for the interests of their members at the expense of non-members. On other hand, most residents in the informal settlements usually do not associate themselves with the municipalities as they do not find the municipal's efforts capable of addressing their water woes after a long period of neglect and seclusion.

PRINCIPLES OF GOVERNANCE IN WATER SUPPLY SYSTEMS

There are three broad principles of governance that can be employed in water supply system. These include efficiency, effectiveness, and trust and engagement. When it comes to good water governance and water security, it should be noted that there is no alternative for water. Water is a finite resource which sustains life and all its activities. It is therefore essential that government organizations prioritize issues of governance and uses that will ensure the future viability and sustainability of water resources. Governance in water supply system should target the following goals; effectiveness, efficiency and trust and engagement.

Effectiveness

The effectiveness of water security governance involves processes and policies that define clear sustainable policy goals and that target all levels of governance. Effectiveness also entails implementation and meeting expected targets. There are four key sub-principles involved. These include setting clear roles and responsibilities; managing water at appropriate scales within an integrated system; encouraging policy coherence across sectors; and, adapting capacity to complex challenges and water sustainability.

When setting clear roles and responsibilities, there several challenges that may hinder the effectiveness of good governance of water security in a place. However, the governance of water security is not managed or implemented centrally or in a unified manner given the structure and the decentralized nature of governance. Fragmentation among various water units may lead to ineffective management of water resources, which adds to the strain of scarce water resources. In a similar vein, studies have revealed that "fragmentation may exist which is likely to cause inefficiencies and ineffectiveness."

On the other sub-principle of water governance (managing water at the appropriate scale or scales within an integrated basic governance system), it should be noted that an integrated approach to water security balances the demand for water from various user sectors and the supply from all the available resources. Because of the diversity in water resources, it becomes very significant to properly manage them in an integrated manner. At the same time, mismanagement of water resources is an important factor that affects water security in a place. One aspect of integrated water management is to have a policy for treated wastewater and how its use can be maximized. Studies have also indicated that an integrated approach to water security may help to addresses issues of emergency conditions and sustainability of desalinated water. There is an urgent need to address this aspect to water security in order to mitigate any risks in an emergency. When doing this, it is important to develop strategic water reserves.

Some governments, especially in those places experiencing severe water security threats like the Middle East, have seen the importance of conserving environment resources as an important foundation to their development policies for present and future generations. Through national legislation that targets environmental issues, these countries have been involved with several regional and international protocols, including the high profile Kyoto Protocol since 2005. Some have even implemented initiatives that encourage long-term sustainability of water resources through Water Conservation Strategy and the Aquifer Storage and Recovery project (ASR). The aim of these initiatives is to actually develop a sufficient capability for domestic water supply to serve all citizens.

The third sub-principle entails encouraging policy coherence across sectors. The effective governance of water security requires coordination across ministries, public agencies, and levels of government. There is the need for integration across different entities in the water sector, to prevent various challenges in achieving policy coherence. There are several federal and local government authorities that are responsible for managing water resources. Therefore, cooperation and support is needed from all the government entities, stakeholders, private authorities, and end users. Even though the economy of most countries is growing exponentially, lack of communication between water entities can lead to poor handling of water resources. This remains a major concern in some countries around the world, especially in developing regions.

Effectiveness involves putting in place mechanisms to encourage coordinated management of supply and demand of water. Some governments are controlling water consumption through price mechanisms such as tariffs. Given that agriculture and green lands continue to be the highest water-consuming sectors, it is important for governments to encourage national agriculture policy aimed at conserving water.

The last sub-principle involves adapting capacity to complex challenges and water sustainability. Building capacity requires innovating and integrating new technologies to improve the effectiveness of water security. There is the need for continuous renovation in order to address the future changes of water consumption.

To address future demand challenges, new technologies need to be incorporated into the development of reliable sources of water supply. For instance, desalination technologies used in some countries such as Middle East are the most advanced technologies available. Due to technological advancements, the costs of producing desalinated water have declined. Most of the Middle East countries have integrated the use of low cost desalination plants. Nevertheless, there is the need for a continuous renovation in order to address the future changes of water consumption.

One of the new initiatives that can be embraced involves developing a water strategic plan. This plan should aim to develop full-scale integrated solution to enhance the water security in a place. It should establish an interconnected system and strategic storage capacity between the major water suppliers in a country.

Efficiency

It is important to consider how a country can maximize the benefits of sustainable water management and welfare with the least cost to society. When discussing governance in water supply system based on efficiency in service delivery, there are three major sub-principles that should be emphasized. These include producing and sharing data and information; mobilizing and allocating water finances efficiently; and effective implementation of regulatory frameworks.

Producing and sharing data and information: One of the key challenges to good governance in water supply system is the lack of reliable data. Data gaps hinder the capacity to produce evidence and research on water security. Data management is an important issue which can be resolved if all the different water institutions in the country function on a single data platform. However, the water governance structure has to be first redesigned by implementing an integrated water resource management approach. Therefore, in order to enhance the efficiency of water security governance, there is the need to emphasize on the effectiveness of water supply institutions and reduce fragmentation. It is important to have a database of all water resources, factors that contribute to water insecurity, water supply activities and consumption. By having one database combining a complex set of variables, the local governments can more easily identify the problem and find a solution for it. Knowledge based resource management backed by intensive research and forecasting will help in reducing the water scarcity issues be it natural or artificial.

Mobilizing and allocating water finances efficiently. According to the UNICEF (2010), there are several ways that governments can address the issue of public cost to water supply production. One approach is through the price mechanism. For instance, some countries like Kenya and Ethiopia have reduced their water consumption as measure to manage the increase in water prices. Other approaches include more efficient desalinated water production through investment in new technologies. According to studies, the use of desalinated water is costly and its use has an impact on the environment. The use of different technology like Reverse Osmosis (RO), electrodialysis and nanofiltration for producing potable water other than the combined power and water technology that are used in desalinated water may be used to reduce water cost and subsequently reduce impact on the environment.

Effective implementation of regulatory frameworks. There is a need to improve the existing regulation and legislation framework in water supply systems. More specifically, the legislation framework has to address the different sources of water such as desalinated water, ground water, and wastewater. Some governments have recommended on the development of common regulations, standards, and specifications for economic, technical and environmental controls in order to address the governance of water security.

Trust and Engagement

When addressing the issue of trust and engagement in water supply system, it is advisable to also look at various steps required to build public assurance and safeguarding the inclusiveness of all the people involved (stakeholders). This should be done in a fair and legitimate manner. There are four key sup-principles that should be considered under the issue of trust and engagement in water supply system. These sub-principles are discussed briefly in the following section.

The first sub-principle entails mainstreaming integrity and transparency practices in water supply systems. The government should enact laws, regulations, and standards that give direction, transparency and clarity in many areas such as the responsibilities, roles, and standards for a particular environment or water sector. However, there is a lack of transparency in water governance standards. For example, in some countries such as in the Middle East, countries are faced with issues on standards for desalination plants that are not clear. Such issues usually emanate from lack of coordination in setting standards. Another key issue involves lack of transparency when it comes to reporting local capital expenditures on various water infrastructure projects. This hinders the collection of relevant data on the cost of water development at the national level.

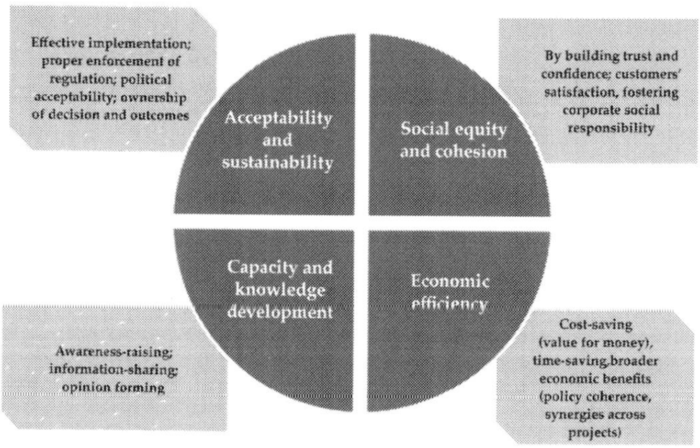

Source: http://www.mdpi.com/2073-4441/8/5/204/htm

Figure 4.8: Stakeholder engagement in water governance

The second sub-principle involves promoting stakeholder engagement. Studies have indicated that water conservation and consumption should be integrated in school curricula. There are several initiatives that educate the public and create awareness about the value of water. For instance, schools should be part of an education campaign to decrease water consumption. The campaign should involve prizes to those schools that are able to decrease their water consumption the most. Another educational campaign involves field visits to water treatment plants. These visits help children gain a better understanding of the cost of water production. These types of awareness building program can help to change behaviour and can bring positive societal change. Educated people are more aware and understands more the need to pay extra money to protect the environment.

Another aspect to water conservation is to ensure that homes have efficient fixtures and appliances that are water efficient. This requires raising awareness among building developers and households to ensure that these types of fixtures are installed. A third area involves promoting the use of treated wastewater. Responsible governmental boards should emphasize that it is important to raise awareness on water reuse, particularly among farmers and the public. Studies show that there is an increase in water consumption in residential (especially in urban) areas. The reasons for the use of expensive water include watering gardens and washing cars. The consumption of water for such uses can be replaced with water of a lower quality, such as treated wastewater. Consumers need to be made aware that there are different types of water and these can be used for different purposes.

The fourth area focuses on encouraging water governance frameworks that help manage trade-offs across users, rural/urban, and generations. It has been stated that in order to improve the governance of water security, trade-offs have to be made across users. For instance, farmers should be encouraged to reduce their use of groundwater and piped water for irrigation by replacing it with treated wastewater. The public needs to be made aware that the use of piped water for washing cars is wasteful and they should be encouraged to consume treated wastewater for this type of use. Government should carry out awareness campaign on regular basis to educate the people on how to effectively use the water. In addition, the farmers should be made aware that the residual minerals in treated waste water can be of great help in growing their crops.

Other trade-offs include raising the price of water on consumers of water to encourage a decrease in consumption. At the same time, there is the need for water management policies for the industrial and agricultural sectors. The WHO (2004) claims that a reduction in the per capita consumption or demand management can solve most water issues.

PROBLEMS FACING INFORMAL SETTLEMENTS IN WATER SUPPLY AND POSSIBLE SOLUTIONS

There are numerous challenges that affect different informal settlement residents in the provision of water supply systems. One of these problems is the poor quality and the high cost of water. Inadequate access to the water sources is also another problem that affects most people. Additionally, the quantity of water is another problem that can be identified as a major challenge affecting most residents in the informal settlements. It is also important to note that most residents are also faced with issues of corruption in the provision of water.

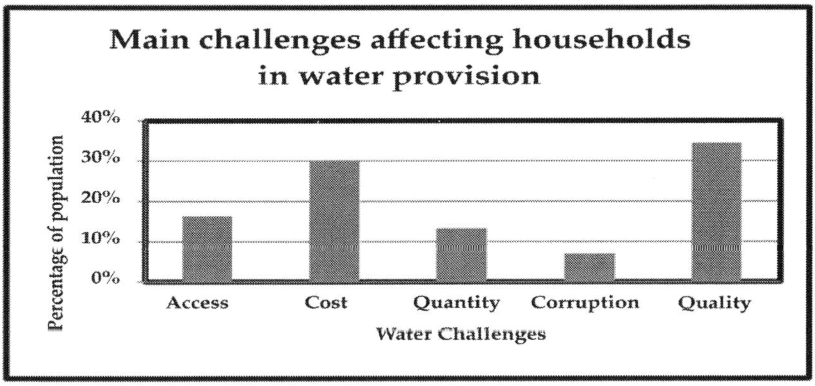

Source: http://slideplayer.com/slide/10826892/

Figure 4.9: Main challenges affecting households in water provision

It terms of preference, most people in the informal settlement would prefer individual piped connections to their premises. However, most of them usually do not trust the municipals to supply and manage the water, yet it has the legal mandate and the infrastructure to provide piped water to the premises. One reason why most people in these settlements do not prefer municipal for their water provision is because of poor record keeping and issuance of incorrect water bills to those who are already metered. In order to deal with water insecurity in the informal settlement, it is significant for the water supply actors to increase the sources of water. Additionally, it also important to focus on the reduction in the cost of water as this would help in dealing with the water problems. Improving the management of the water points would also help to alleviate the present water crisis. Another proposal towards solving various challenges experienced in the informal settlement with regards to water provision is repairing water pipes.

***Challenges Facing Water Vendors**:* Most of the challenges faced by water vendors may vary depending on whether they are experienced in obtaining water or when selling water. Most of water vendors in the informal settlements are faced with the problem of harassment by authorities. This is actually one of the greatest challenges when obtaining water. The authorities involved included officials from the water company, the officers from the municipal council and the local administration police. Another key challenge that the water vendors face are the leakages from pipes, especially when there is water shortage. Poor quality of water and high water prices are also some of the important challenges the water vendors face in their day to day water provision in the informal settlements. Poor roads are also a major impediment in the quest to sell water. Cumulatively, over 80% of these challenges are not attributed to water shortage but rather to the lack of a proper water governance structure and failure of the relevant institutions to carry out their respective mandates.

Despite the challenges faced by water vendors in obtaining and selling water, there are several proposals that can be implemented in order to improve the situation in the informal settlement. Improving the infrastructure including paving of the roads and repair of leaking pipes will really help in enhancing water provision in the informal settlements. This is however a challenge, given that there is hardly adequate space to enhance accessibility and movement in the informal settlement. An end to the constant harassment by the various authorities will also improve water provision by the vendors in the informal settlements. Other key proposals that can be recommended include increment of water kiosks to increase the number of water outlets and increasing individual water connections for the residents.

Challenges Facing Municipalities in Water Provision: Most municipalities in charge with water provision in various urban areas usually obtain their water from the surrounding peri-urban areas. The adequacy of water in these areas dictates how much water is available for city residents, including those people living in the informal sentiments. Studies have shown that dwindling rainfall in the various water catchment area and increased abstraction rates usually result to less water available to cope with the high demand in most urban areas.

The quest to quench the rapidly increasing water demand by exploring alternative sources of water such as underground water and rainwater harvesting has not been fully exploited in most places around the world. There is very little empirical evidence of this, if any. In most places, underground water is not fossil, but renewable. However, the rate of extraction is quickly exceeding the rate of recharge with new evidence emerging of dried up boreholes due to exhaustion of the aquifer. The capital and maintenance costs associated with drilling boreholes are huge. Water service provider in different countries are embracing the idea of harvesting rainwater as a viable option that will ease the demand of water services for non-consumptive uses.

Some places around the world have uneven topography, and this usually acts a major hindrance in the efficient flow of water to major parts of the area. Another key challenge that municipal companies may face in the provision of water in a city is lack of adequate pumping and distribution stations compared to the steep water demand in most cities, which continues to grow both spatially and demographically.

Rising energy costs are a major impediment to most municipalities' quest to supply water. Most municipal companies usually run their water pumping stations using diesel powered generators and electricity from the national grid. Oil prices which are dictated by global prices have been on a steady increase in the recent past, thereby significantly increasing the operational and maintenance costs of most water companies. These costs are rarely levied on consumers and with no subsidy from the government, the company foresees passing the burden to the end users if this steady rise continues. According to recent water survey conducted by World Health Organisation in Western African countries, ineffective revenue collection mechanism was identified as a key challenge that results in less than half of the consumers connected to the grid system paying their bills. Therefore, responsible water companies should embrace effective mechanisms such as Information and Communication Technology (ICT) where customers can check their outstanding bills via text message and pay using mobile money services.

Leakages and burst pipes and blocked sewers remain a major problem. The aging and frail water infrastructure in most cities can no longer adequately serve the city. Most pipes in water supply network have failed from over-use and lack of maintenance over the years. Most water companies in different countries have also been battling with illegal connections since its inception. This is a very common and highly prevalent practice especially in informal settlements where many cases of vandalism go unchecked and unpunished. Water cartels sell illegally diverted water from the main water piped network to poor urban dwellers at unreasonably high prices. One of the best ways the responsible water companies can embrace with accordance with this challenge is to involve the affected informal settlements in constructing more water kiosks and providing more storage tanks in areas without the piped network.

Source: https://slumurbanism.wordpress.com/category/slum-upgrading/page/4/

Figure 4.10: A typical illegal water connection in the informal settlements

Additionally, all residents in the informal settlements should be encouraged to report incidences of vandalism. This initiative may be strengthened through having an effective legal framework and an adequate capacity to enforce the law. The municipal company in charge of water provision in the informal settlements should conduct civic education programmes. These programmes should target both domestic and industrial users. They should offer a platform for interacting with the communities and advocating for water conservation tips. Some of residents in the informal settlements are involved in reusing water for non-consumptive uses like car washing among others. Municipal companies should also use these educative programmes to explain the rationale for their water rationing and also get feedback from households on how to improve water services.

5 FINANCIAL EVALUATION ON SMALL-SCALE WATER SUPPLY INVESTMENTS

INTRODUCTION

Finances for investing in a certain project are provided to the company by investors who hold various types of claims on the company's cash flows. The optimal investment decision maximizes the expected satisfaction (expected utility) gained from consumption over the planning horizon of the decision maker. Sound financial management and capital investment decision making are critical to survival and long term success of a water utility company.

After identifying a viable project, management should then begin the financial process of determining if the project should be pursued. According to Discounted Cash Flow (DCF) methods, any project should be evaluated thoroughly to ascertain its feasibility to getting financial help from a reputable source. Discounted Cash Flow (DCF) methods which takes into consideration the time value of money, is regarded as theoretically correct and includes at least three different discounting models which are:

- The payback periods
- Net present value (NPV)

- Internal rate of return (IRR)

The payback period basically determines how long it takes to pay back the initial investment that is required to undergo a project. The payback period is probably best served when dealing with small and simple investment projects. If the business is generating good returns, it is supposed to recoup its investment in a few years. Net present value (NPV) on the other hand considers the time value of money, because the cash flows are discounted back at the firm's rate of capital (r). This rate is the minimum return a firm must earn on a project to have the firm's market value remains unchanged. If the amount earned on the project exceeds the cost of capital, NPV is positive, so the project adds value and should be accepted. The larger the NPV, the more financial value the project adds to the owner (s).

Internal rate of return (IRR) is essentially the return to be received over the life of an investment. It is calculated as the discount rate at which NPV equals zero. This is the rate that makes the present value of the cash flows equal to the initial investment. Strictly defined, the internal rate of return is the discount rate that occurs when a project is at break even, or when the NPV equals 0. The decision rule is simple (in that, you should choose the project where the IRR is higher than the cost of financing). It is also worth noting that the greater the difference between the financing cost and the IRR, the more attractive the project becomes. Each project that uses internal funds has a cost of capital.

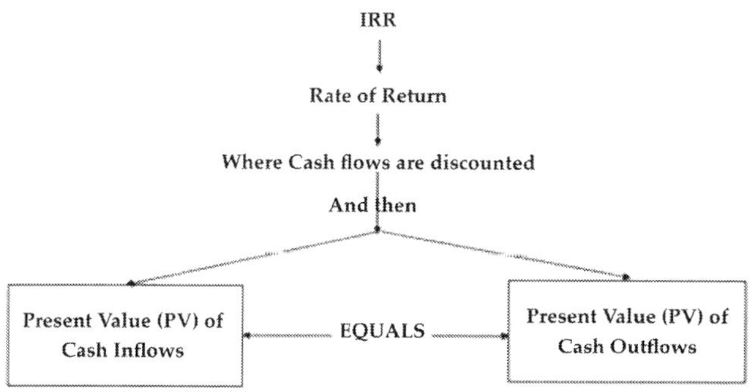

Source: https://www.mpug.com/articles/pmp-prep-understanding-internal-rate-of-return/

Figure 5.1: Internal rate of return matrix

Internal rate of return should thus be compared with the cost of capital. If the rate earned is more than the rate it costs, then the project should be undertaken as it adds to firm's value. The goal with capital budgeting is to select the projects that bring the most value to the firm.

The World Health Organization (2010) emphasized that water is one of the most basic requirements for human existence, yet over a billion people in the world lack access to it. Most of the utility companies involved in water supply are usually managed publicly. Besides the revenue collected from distribution and provision of water, most of these companies depend mostly on the governmental financial support and subsidies. Nevertheless, public water utility companies in most parts of the world have been unable to offer universal access to water services because of financial constraints. This has negatively affected the level of investments by public utility companies in water sector, especially in urban and peri-urban areas.

Source: https://www.youtube.com/watch?v=brw7ghc4Izc

Figure 5.2: Illustrating a typical Small-Scale Water Provider (SSWP)

The rise of Small-Scale Water Providers (SSWPs) reflects the inability of public and private water utility companies to adequately provide for the water needs of city dwellers. For instance, in a country like Ethiopia, millions of people, especially the urban poor, remain underserved or not served at all by public or private water utilities. Due to insufficient structures coupled with rapid population growth and urbanization, the gap between demand and supply of water continues to widen. Increasingly, SSWPs are being acknowledged by different governments around the world as vital suppliers in water sector. However, World Bank has revealed that financial constraints faced by these SSWPs, hinders their growth and expansion.

Most water utility companies usually report negative incomes because most of them usually set the users' fee below full cost recovery level. According to studies, low cost recovery is often regarded as a major contributory factor to the poor sustainability of urban water management in developing countries. In other words, it has been found to negatively affect the capability of utility companies to enhance water provision services coverage. Because of this impact, most water supply systems in both urban and peri-urban areas cannot thrive without financial assistance. Researchers have also revealed that low cost recovery usually make economic viability even more difficult.

Expanding the existing water infrastructure has become a nightmare as the risk of investing in most countries is too high. Whether funding can be secured from the financial system (and at what costs) will be determined in large part by the risks that investors think the water project entails. According to World Bank (2010), the level of risk is usually estimated by looking at the experience investors have had with similar projects in the past. Volatility of water investment makes most investor fear the risks involved. Even if lenders are willing to finance the project, they may charge a high risk premium which pushes up the cost of financing. For some water projects to be financed, external group assume part of the credit risks.

Going with the statement given by the World Bank (2010), it has been established that water investments are by their nature capital intensive and yet this capital is not readily available. Private players are not willing to borrow to finance water investments due to the predatory interest rate and high cost of capital. The creditworthiness of a water project can be judged based on availability of assets pledged by the borrower rather than a borrower‘s expected revenues and cash flows. Lenders are normally cautious and are likely to require collateral that can be taken in the event of default. Small Scale Water Providers often must satisfy collateral requirements well more than 100 percent of the loan amount needed.

Most water investments are characterized by low return on investments as different governments are usually not willing to improve/raise water prices to a level that meets the market threshold. Most of the cash that is generated through the sale and distribution of water is to cater for various services such as paying for the operating expenses, maintenance expenses, debt services and profits to the owners of the project. Therefore, in case of an investment with low returns, the utility company is usually faced with a lot of difficulties in covering the financial gaps experienced. As suggested by the World Bank (2010), one of the best ways to increase revenue generation by the water utility companies is to charge water users directly the full costs of water services offered to them.

GLOBAL TRENDS IN WATER INVESTMENT

Urbanization and population growth in most parts of the world has been increasing rapidly, thus posing numerous challenges to different governments. One of the most challenges that governments face in various countries is providing safe and clean water to the citizens. Population growth rate and urbanisation are contributing to the straining of the scarce water resources in most urban areas around the world. The UNICEF (2013) stressed that the provision of safe drinking water contributes to sustainable improvements in people‘s lives regarding their health, education and economic situation, eradication of extreme hunger and the empowerment of women. In other words, when providing water services, there is need to balance the social and economic needs for water. This makes managing urban water a challenge, especially given that the demands for water are increasing yet the availability of the resource is decreasing mainly due to financial constraints.

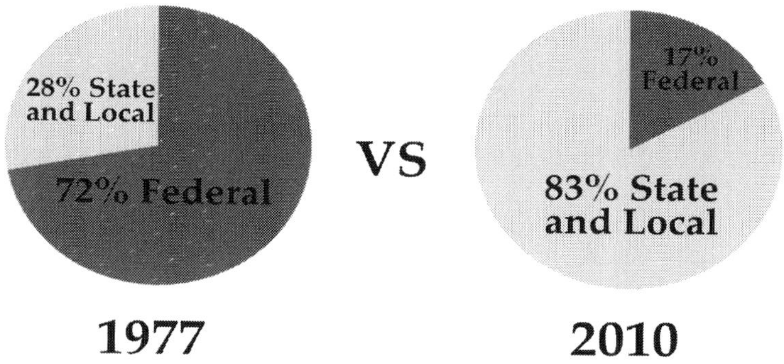

Source: http://www.circleofblue.org/2014/world/u-s-water-systems-deteriorated-slow-change-need-new-strategy-money/

Figure 5.3: Capital investment in water infrastructures in U.S.A.

For instance, in United State of America, an investment in water provision in the year 2007 was worth about $20 billion, catering for about 240 million citizens. The World Water Council (2007) stated that some of the key challenges facing various water users especially in the urban centres in U.S.A include water scarcity, water quality, and water affordability.

In a country like Malaysia, water supply investment had been a responsibility of the 13 states. Between 2001- 2005, federal government had allocated US$ 1.1billion for water supply projects. However, faced with US$ 2.2 billion debt by state utilities, the federal government decided to embark on a sector reform. Several states of Malaysia including Kuala Lumpur embarked on a policy of private sector participation in water provision committing to invest US$ 760 million.

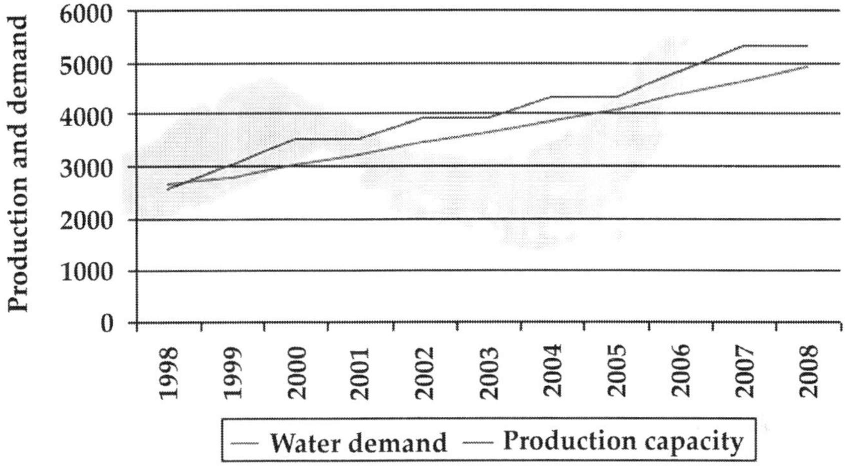

Source: https://www.slideshare.net/iwlpcu/integrated-water-resources-management-in-malaysia

Figure 5.4: Production capacity and water demand (MLD) in Kuala Lumpur and Selangor

This was followed by US$ 2.5 billion investment in 2004. According to World Bank (2010), the result was a mixed reaction in the sub-sectors as it didn't works as expected. In 2009, Malaysian utilities managed to recover on average, 15 percent of their operating costs. But it should be kept in mind that operating costs were kept low through various subsidies. The full costs of service provision operating costs plus capital costs were not covered through revenues.

Another good example worth looking at is the water provision system in India. This country has a heterogeneity structure in water supply systems. About 70% of water is supplied through piped water to various household in big urban areas; another 45% of piped water is supplied in small towns, while only 9% of households in the rural areas are provided with piped water. The remaining percentages are served by the private and small-scale water providers.

Between 2007 and 2012 Indian government had planned to invest US$ 24 billion or US$ 4.8 billion per year for urban water supply. The central government financed 55 percent of the investments, 28 percent was financed by state governments, 8 percent by the institutional financing, 8 percent by the external agencies and 1.5 percent by the private sector. The volume of investments was expected to double to reach 0.7 percent of GDP by 2012. Government subsidies in India account for 4 percent. About 98 percent of this subsidy is said to come from state rather than central budgets.

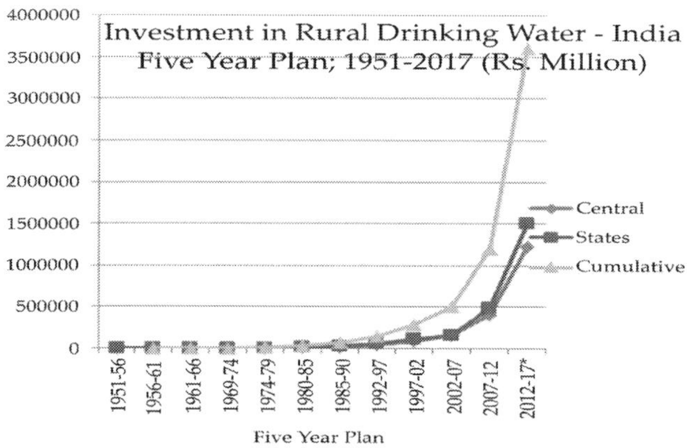

Source: http://catechesis.net/?mqg=invest-in-water-india

Figure 5.5: Illustrating investments in rural drinking water in India

The state and federal governments in India spend US $1.1 billion or 0.5 percent of GDP in subsidizing water between 2007 and 2012. Water investments that were not accounted for were estimated at 25 percent to 40 percent of water produced by utilities in the main urban areas in India. Overstaffing in Chennai and Delhi is as high as 33 per 1000 connections. Under-pricing made most of the utilities to be unable to recover their costs. The average rate of cost recovery in 20 cities in 2007 was 60 percent. The main challenges affecting Indian government in water provision includes un-accounted water, overstaffing and low cost recovery mainly due to under-pricing of water.

According to Public Utility Board of Singapore, water supply in Singapore is characterized by a number of achievements, among them is the universal access to water that is affordable, efficient and of high quality. In the financial year 2010, Public Utility Board (PUB) in Singapore undertook investments of US$ 290 million in its water infrastructure. In the year 2005, PUB issued for the first time a bond, raising it US$ 400 million to finance part of its investment program.

During the financial year 2010, Public Utility Board group received an operating grant of US$ 185 million to fund its operation. Water tariffs in Singapore are set at a level allowing cost recovery, including capital costs. The tariffs were raised in the late 1990s from US$ 13 in 1996 to US$ 30 in 2000. By the year 2012, water tariff included a conservation tax set at 45 percent for domestic consumption above 40 cubic meters per month. A general service tax of 7 percent is added to the bill. According to Public Utilities Board, industrial water tariffs are set lower at S$ 0.52 per cubic meter.

According to Human Development Report (2006), the annual investment needs in water supply in Bangladesh are estimated at US$ 150 million. Actual annual investment needs in water in urban areas have been estimated at around US$ 40 million per year though only US$17 million per year is invested. External funding accounted for 96 percent of this investment in the sector. Community Water and Sanitation Agency (CWSA) maintain that the public water utilities are responsible for providing, distributing and conserving water. Local private companies are in charge of meter installation, customer billing and revenue collection.

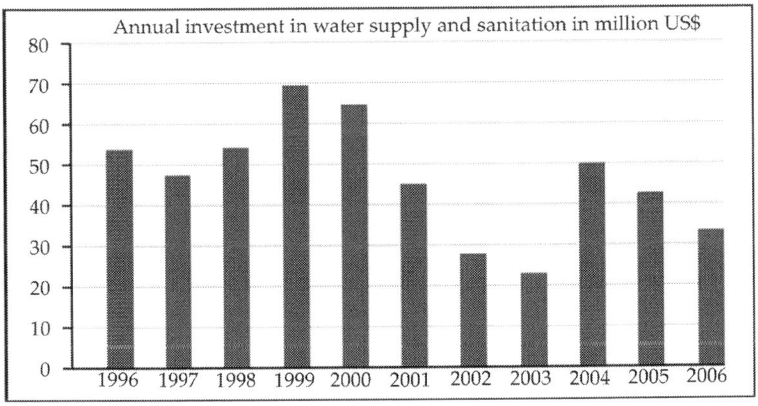

Source: https://en.wikipedia.org/wiki/Water_supply_and_sanitation_in_Bangladesh

Figure 5.6: Estimates of annual investments in water supply in Bangladesh

Based on the viewpoint given by Water Aid (2011), water supply in Bangladesh faces several challenges including very limited access to water, non-continuous water supply, high water losses and low water pressure. Non-revenue water in urban areas stands at approximately 50 percent of the produced water. Illegal connections account to 3,000 out of 15,000 connections. Most of those connected to water supply do not pay their bills. At the end of the 1990s, the Bangladesh government participated in that poor payment culture. The year 2010 reported low labour productivity of 7.2 persons per 1,000 connections. Water tariffs in Bangladesh were too low (US$ 0.15 per cubic meter) to recover the costs of the service. According to World Health Organization (2010), access to water supply in Ethiopia is amongst the lowest in Sub-Saharan Africa and the entire world. The government estimated that the actual investment needs are about US$ 297 million per year for the period 2006-2015 but the actual investments was approximately US$ 39 million. This shows a very large deficit. Investments for 2006-2015 are estimated that 12 percent was funded by the government, 15 percent by communities and 73 percent by donors.

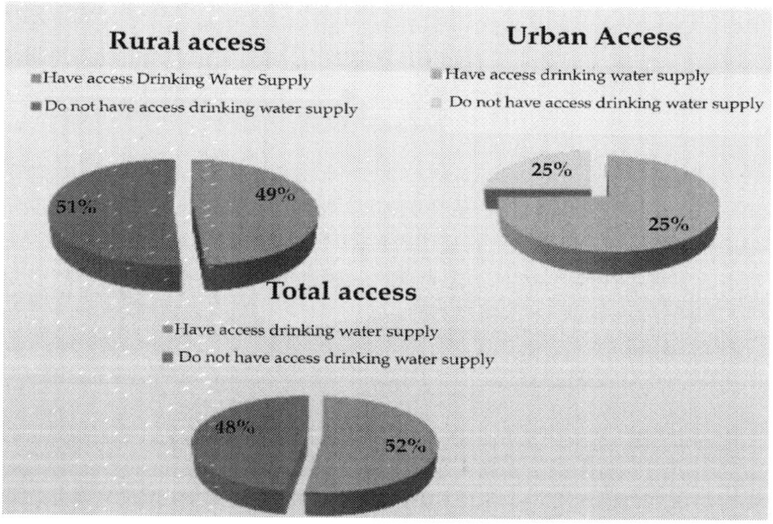

Source: https://ethiopiaobservatory.com/2013/07/15/ethiopias-drinking-water-coverage-drops-by-17-sign-of-a-more-serious-problem-part-i/

Figure 5.7: Water supply in Ethiopia

Water in Ethiopia is supplied by both utility companies and private operators. Ministry of finance and economic development states that the year 2011-2015 is aimed at increasing drinking water coverage from 68.5 percent to 98.5 percent. World Health Organization (2010) indicated that only 38 percent of urban population had access to improved water supply. Few service providers recover all operating costs and generate cash surplus. Non-revenue water, low labour productivity of 9 persons per 1000 connections and illegal connections are the main issues in Ethiopia.

Due to challenges and investment condition that characterized most public water utilities, many governments have taken the initiative of exploring increased private investments. They are trying to expand their access to new financial resources, technical and managerial skills. However private players are reluctant. No private company has provided effective water services to label any of the privately provided water service projects a full success. In recent years, cases in Buenos Aires in Argentina, Manila in the Philippines, Atlanta, Georgia, Cochabamba in Bolivia, Jarkata, Indonesia, United Kingdom and South Africa, indicate that water privatization in most countries have suffered major losses. This according to World Bank (2011) has made peri-urban population to buy water from SSWPs.

FACTOR THAT INFLUENCE WATER SUPPLY FINANCING IN A COUNTRY

In any water supply project, there are several factors that influence financing process. These factors can be classified as either dependent factors or independent factors. According to studies, it has been revealed that most independent factors usually cause changes in the dependent factors. Some of the independent factors that may influence financing a water supply investment include cost recovery, perceived risks of the investors, capital access and return on water supply investment. The following figure outline key dependent and independent factors that may influence financing water supply investment.

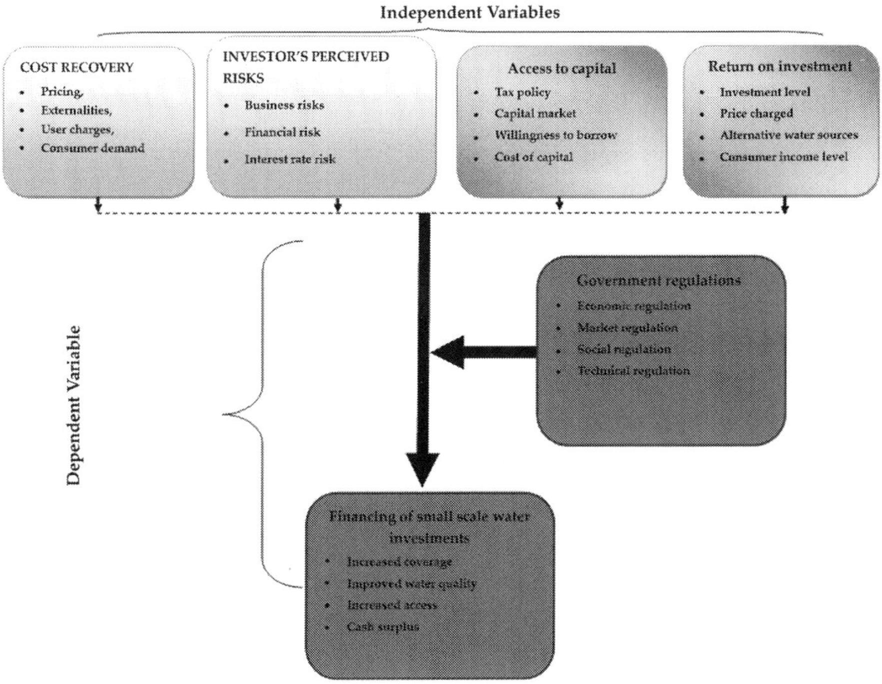

Source: https://www.urban.org/research/publication/managing-investors-risk-pay-success-projects

Figure 5.8: Dependent and independent factors that may influence financing water supply investment.

Cost Recovery

According to UNDP (2003), cost is the major limiting factor of access to clean water. High costs, low efficiency and unreliability are the characteristics of many utilities in developing countries, especially in rural and sub-urban areas. The financing cost of water utilities exceeds the capacities of the public sector utilities. The cost to replace the deteriorating water infrastructure in industrialised countries may be as high as $ 200 billion a year. The type of cost, pricing, consumer demand, external conditions and charging users directly are vital issue for the full cost recovery.

It is actually difficult to determine the optimal price because of the absence of competitive areas in water supply. Step tariffs encourage water efficiency use, as the marginal price of water increases from the first to subsequent blocks.

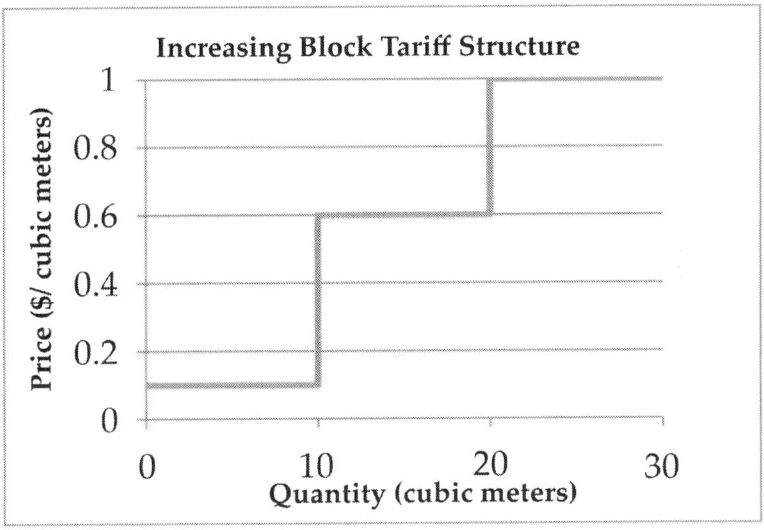

Source: http://www.sswm.info/content/water-pricing-increasing-block-tariffs

Figure 5.9: Water pricing increasing blocks tariffs

One of the key aspects for the provision of water service in any place around the world is affordability. This affordability of water supply is directly related to water pricing policies. It is worth noting that this affordability of water supply service may not equally be distributed across all the income groups in a society. In other words, those who are in the lower income group are inevitably likely to pay a higher proportion of their income in order to have access to clean and safe water compared to those who are in the higher social-income stages, and thus it is important to consider the social and economic value of water. Major development agencies such as the World Bank support the economic concept of willingness to pay for water for full cost recovery.

According to World Bank (2005), charging water users directly to recover the costs for water service may result in a decrease in water use and facilitate the private provision of water services. According the argument of World Bank (2005), charging customers the full cost for service delivery is necessary but unrealistic, because many people cannot afford to pay the full cost of water services. The preferred method is cross-subsidization, which means charging the wealthy more than cost and the poor less than cost for water services instead of using public funds. It is important to promote a long-term target of full cost recovery, but the process needs to happen slowly and citizens need to see improvement in services in order to be willing to increase payment for water services. All developing countries should be encouraged to implement direct water usage fees, which later may enable the private provision of water services. Cost recovery can also be implemented by public water utilities to promote the efficient use of water.

The plea for water service demand for services is shown in prices. This in other words means that as demand increases people's willingness to pay more also increase and this consequently improves the service delivery of a water utility. The excess price paid over and above the prices that justify the true value of a product may be indicators of consumers demand for product or service. Service demand is explained in terms of natural market failure due to economic conditions like inflation, tax and interest rates. These elements influence the relevant prices of factors of production which in turn have an effect on demand for services. Demand for water is affected by price charged, quality of water, consumer's income and alternative water sources.

There are different factors that influence willingness to pay for service offered in terms of water utility fees. These factors can be described using the linear population model below:

$$WTP_i = \sum_{i1}^{n} \beta_i X_i + \varepsilon_i$$

Where;

X = represents a vector of independent variables that may influence an individual's willingness to pay for water supply service deliver through the water utility bill.

B = represents a vector of parameters to be estimated, and is normally distributed random error term with mean zero and constant variance. The expected willingness to pay of an individual is represented by the coefficient *i.Please define ε also.*

Perceived Risks

Studies have shown that risk and economic activity are inseparable. In other words, any investment involves a trade-off between risks and return. Most of the risks that can be identified in water supply investment may come in different forms; business risk, financial risks, interest rate risk, and liquidity risk among others. Studies have also shown that a commodity price risk is the risk of market price volatility. In case of fluctuations on water supply system, there can be huge disparity in terms of the costs of procuring water supply services (can either be high or low) compared to what was anticipated. This may in turn result to either increase or decrease in profits gain. There are numerous alternative water sources that can be used to create a commodity risk to an investor. In particular, increased water prices and other related regulations are a key outcome of recent policy changes in the water sector. Such policy pressures the condition of water consumers to regard water as increasingly costly and unreliable and so motivate them to minimise the amount of water that they consume. A project is bankable in the private sector if it is financially viable and sufficiently robust to survive a downside risk scenario without the financiers losing their money. If either of these requirements is not met, private finance will not be forthcoming and the project will have to be developed using public funding.

Access to Capital

Capitals provided for investment come from variety of sources including bank loans or mixed systems. An efficient credit market will support competition between different types of lending.

When accessing the capital for water investment, developing new financial products or providing guarantees will not help if private investors and water utility companies are unwilling or unable to borrow mainly due to prevailing economic environment. There are several factors that influence the willingness of an investor to borrow cash. These factors include tax rate, inflation and interest rates.

From the private or small-scale water projects perspective, borrowing may be an option if citizens demand better services and if they are able and willing to pay for them. Removing barriers on the demand side is a pre-requisite to developing active and efficient credit systems. Many governments have created Municipal or Urban Development Funds to channel credit to water utilities for investment. In much of the cases, these funds have been established in cooperation with international organizations like the World Bank, regional development banks or bilateral donors.

Governments with many pressing and competing commitments for budgetary transfers cannot be relied on entirely to financially support water utility operations, left alone capital investments needs requirements. According to World Bank (2005), funding may be available but more often on a sporadic or non-recurring basis. Often, these are rationed to accommodate several competing needs across different sectors and purposes.

In most parts of the world, the responsible water authorities and the private operators have failed to extend piped water to parts of the town due to lack of funds. A view is gaining ground amongst lenders, donor agencies and other observers that there is a low level of investment in the sector and as a result little demands for finance other than for grants. This is because of high cost of capital that is inaccessible.

Return on Investment

Most of the countries around the world are still experiencing financial challenges in infrastructures for water supply system, this is despite of the the United Nations' declaration of 1980s as International Decade for Clean Water and Sanitation and the increased funding for water development by International Financial Institutions (IFIs) and national governments. This is mainly where water infrastructures are financed by public utility companies.

The World Bank (2006) suggested that there is the need to advocate for charging water users directly to recover the full costs of service provision (instead of subsidizing delivery through public taxes). This will help mainly in facilitating the transition from public to private water service provision and to increase revenue generated. World Bank (2004) stated that involvement of private partners in project ownership separate the policy-makers from water service providers and makes providers more responsive to clients. This would lead to increased focus on customers, improved customer relations, improved quality of water and increased revenue for the business.

Recently, most countries around the world have begun privatizing their water utility services. However, most water supply services are still run publicly since different private water utility companies are uninterested in managing water supply utility services that generate low returns. There is little or no profit at all to be made due to low water prices as government are unwilling to raise prices to market levels.

The huge investments required for increasing water supply service distribution result in a marked lack of incentive for private companies to invest in the least wealthy areas because they are unprofitable. The result is that private investment is not going where it is needed, most that is, to those without access to clean water or invest in the least wealthy areas because they are unprofitable.

Low income users cannot pay the full costs for the service required for the company to maximize its returns. Ability and willingness of a household to pay for water depends on many factors, among them, the household income and quality of service delivered by the service provider. However, in developing countries, due to the poor service delivery by most water utilities, water shortages are prevalent and this leads to many residents to resort to unsafe water sources.

FACTORS THAT INFLUENCE WATER SUPPLY FINANCING IN A COUNTRY

A pressing issue for maintenance and expansion in most developing countries is adequate cost recovery and though most countries in the world have national water policies and programs, none of the cost recovery strategies consider the replacement or rehabilitation cost of fixed assets. An objective of a country's National Water Policy for urban supply should be to ensure a financially viable utility and adequate funding level to rehabilitate, improve and expand infrastructure, and to undertake operation and maintenance. Though some measures have been put in place for achieving this target, most countries, especially in developing countries still have a long way to go in enforcing these measures to derive their full benefit.

According to World Bank (2005), financing water infrastructure means spending money to finance long-term physical assets and sustainable financing implies that expenditures are balanced with revenues. Most countries in the world have developed or are in the process of developing financing strategies in water supply and sanitation. A challenge in long term infrastructure development is the overriding political nature of decisions on infrastructure investments considering the short term nature of the political cycle. To ensure proper asset management, one needs to secure adequate long term investment funding for maintenance and renewal.

According to a report by the World Bank (2004), there are two primary approaches to financing water system improvements and maintenance: cash financing or debt financing. Cash financing is usually limited to the revenue at hand, which is from water rates, service fees, connection fees from new accounts, or taxes. Debt financing is the typical way that water utility companies raise upfront capital to invest in their systems and it includes collecting loans to finance the infrastructure. There are three main sources of funds for developmental projects in water sector, namely

- External support agencies,
- Government's annual budget and other development votes
- Internally generated funds mainly through water sales

American Water Works Association (AWWA) is committed to the principle that utilities should be self-sustaining through their rates. This association recognises that huge investment is needed by the water utilities to maintain and replace existing infrastructure and wonders if this can be done by the utilities financing themselves at tariffs customers can afford. Many scholars consider tariff revenue as the most important source for cash flows and charges should contribute to investment costs as well as finance maintenance and operation. According to UNICEF report of 2010, a self-sustainable tariff should cover the operation and maintenance of the system, investment for the rehabilitation and renewal of the distribution, investment in infrastructural development and return on capital employed. A mandate for water and sanitation utility companies, especially in developing countries, is cost recovery and about 60% of the utilities are meeting their full operation and maintenance cost. Cost recovery through tariffs is also a challenge in most countries.

Another source of funding within the water sector is grants from donors and commercial funding. For instance, a country like Kenya has been receiving donor investments in the water sector. The general lending trend in the water sector may vary from time to time based on the economy of a country.

MAINTENANCE COST OF WATER SUPPLY SYSTEM ASSETS

There numerous services are needed other than just constructing the infrastructures for water supply system. For instance, there is the need for operational services, maintenance services and augmentation of various infrastructures of water supply systems. Maintenance can be either planned or unplanned. It is planned when it is organised and carried out with fore thought and unplanned when there is no predetermined plan and is necessitated by breakdowns.

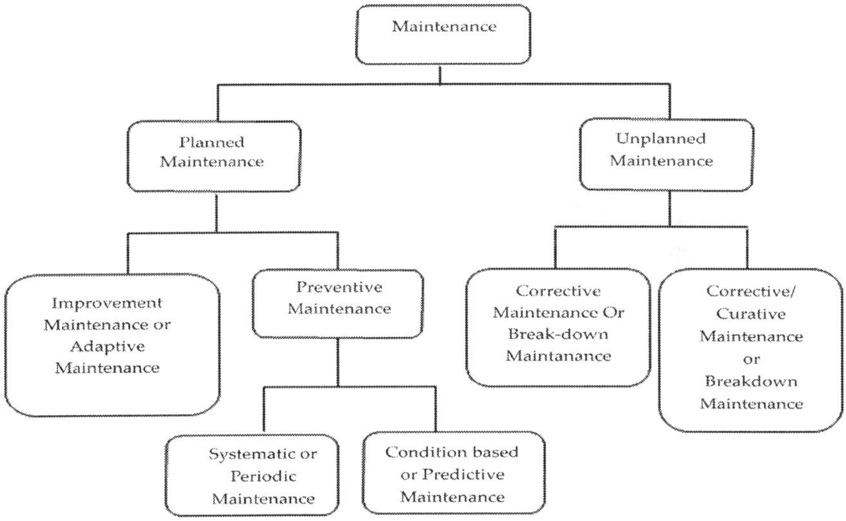

Source: https://www.omicsonline.org/open-access/hotel-maintenance-management-practices-2169-0286-1000136.php?aid=78487&view=mobile

Figure 5.10: Classifications of maintenance in water supply system

There are different strategies that are employed in maintenance. The most common strategies used include preventive maintenance, breakdown maintenance, predictive maintenance and proactive maintenance. Total maintenance cost is presented as the sum of the direct costs of maintenance activities, the cost of breakdowns, failures and stoppages of production equipment, and the cost of writing-off production equipment. Some of the losses that are usually associated with failure include direct costs, indirect cost and social costs.

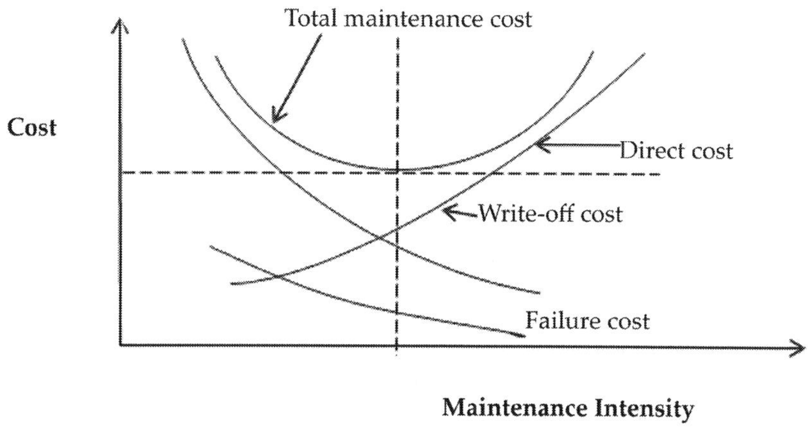

Source: http://www.risktec.tuv.com/knowledge-bank/technical-articles/emit-optimisation---getting-more-out-of-existing-equipment-for-less.aspx

Figure 5.11: Optimisation of maintenance cost

This depicts that as the maintenance effort in water supply system is increased (do more preventive maintenance), failure costs and write-off costs will decrease. According to World Bank (2010), the "rate at which capital depreciates depends positively on its utilisation and negatively on maintenance expenditures." It is important to consider the fact that more maintenance will produce less failures and longer lasting equipment as a narrow perspective of maintenance. A recent form of maintenance that is becoming popular is Reliability Centred Maintenance (RCM) also known as Risk Based Maintenance (RBM). The aim of RCM is not to preserve the assets in themselves, but to preserve the asset's functions (whatever the user wants it to do, as defined by performance standards). Thus it does not aim at preventing failures but rather preventing the consequences of failure. Risk-Centred Maintenance originated from the realisation that more maintenance does not automatically translate to less stoppage or failure. More maintenance especially preventive maintenance may even cause failure rather than prevent it.

There are different options for the maintenance of a water supply system. Most of these options usually deal with the actual assets over time. They are:

- Operate and maintain the existing asset
- Repair the assets as they fail-this aims at reviving an asset to bring it back to its original state
- Rehabilitate the assets- rehabilitation brings the assets back to a functional condition and may extend the assets' life span considerably. It refers to a one-time event when a major activity is performed on the asset to extend its life so that the asset continues providing service for some additional period of time
- Replace the assets- at a point in time; most assets in water supply system will require capital for replacing them, especially if they have been in services delivery for a longer period.

Most of these options are closely linked to each other. The frequency with which one option is used impacts the frequency of the other, if the other is done at all, or the time frame in it is done. Some infrastructure assets in water supply systems can remain in the "operation and maintenance phase" for a long time with very little probability of decommissioning. However, to maintain high levels of water supply services delivery, there is the need to occasionally carry out a system upgrading or improvement services.

COSTS FOR WATER SUPPLY SYSTEM

Simply, cost can be defined as "*a measurement, in monetary terms, of the amount of resources used for production of good or rendering services.*" The costs of water supply system can be classified in to several concepts. These concepts are discussed below.

Direct Costs: These are costs that are usually traceable exclusively to an output and relate to provision of the service offered.

Indirect Costs: Are the costs that are not exclusive to a particular output and are often relatively insensitive to changes in quantity of output.

Fixed Costs. These are the cost that are usually constant (in other words they do not vary) based on the water quantity that is produced.

Fixed costs for a water supply system include administrative costs, scheduled maintenance and equipment replacement costs.

Variable Costs*:* These are costs that are directly influenced or dependent on amount of water volume that is discharged. These costs include energy costs, chemical costs, and cost of raw materials. Sometimes it is difficult to categorise some cost components because they can be variable or fixed depending on the situation, e.g. labour and some forms of maintenance.

Unit Cost*:* A unit cost is the cost of producing one unit of a given service. Since unit costs relate input to output, they are good management tools. It gives a relationship between the resources consumed and the resulting product.

Recurrent Cost*:* These are costs that are incurred in various process of water supply system. These costs include maintenance expenditures and the operations expenditures. These costs are important because they help in running a water project or supply system at a level consistent with its expected use and to maintain its capacity during its expected lifetime.

WATER SUPPLY MAINTENANCE EXPENDITURE

Costs of maintaining water supply system are expenditures which are usually related to preserve the system (for instance, the costs for repairing water supply assets). In other words, the expenditure incurred in maintaining assets is termed as maintenance expenditure. Capital maintenance is an important element of water supply system. In other words, "*Capital maintenance is planned work carried out by companies to replace and refurbish water and sewerage assets to provide continuing services to customers.*"

Since water supply systems consists of different assets with different lifespans which wear out at different times, there is the need to have a system that will ensure that funding for the water supply system is in place to ensure continuing water supply services. Without capital maintenance, many water supply systems may be abandoned and the full benefit that should be derived from the infrastructures is usually not obtained. Studies have shown that a failure of serviceability in a water supply system may lead to a very serious public inconvenience.

It has been stated occasionally that the true cost of water delivery cannot be obtained without considering the cost of capital maintenance expenditures. This is because as the assets are used in producing and supplying water, they depreciate and will have to be maintained to restore them to acceptable levels of performance. The cost of maintaining the assets must be known and provision must be made for it so that assets will not fail and water service delivery will be sustained.

There are many ways to estimate the capital maintenance for water assets. Some people use depreciation of assets over their useful lives and thus determine the amount needed to restore the assets. The Office of Water Services (OFWAT), a regulatory body for water companies in the UK determines the capital maintenance for water utility companies based on historic levels of capital maintenance, the trends of service to customers and the scope for further capital efficiency both catch-up and minimum annual improvement. Some of the water companies argue that this is not an accurate basis for the estimation of capital maintenance as times and conditions have changed. In using historical prices, necessary price adjustments must be made to reflect changing condition of the local environment. In the recent practice, a price index is usually used if historical data on capital maintenance will be used as a basis.

A challenge in determining the capital maintenance of water supply systems is the unavailability of data. This is common for new system, systems where major maintenance has been neglected or where there is no proper record keeping.

Most water utility companies usually have a maintenance policy, which considers routine maintenance of various assets and minor replacement of parts. The maintenance policy, however, does not consider capital maintenance of the systems. It has been revealed that maintenance at the water supply systems is usually done in two main forms namely; (1) Preventive Maintenance (2) Breakdown maintenance

Preventive Maintenance*:* These are regular maintenance carried out on components of the water supply system such as pipes, reservoirs, civil structures, pumps, and other electromechanical equipment to ensure that they function according to design. Well-structured preventive maintenance programmes with accompanying schedules are adhered to rigidly. Operational manuals from manufacturers are normally re-written in a simplified form for use by the technical operators of the system. Maintenance schedules in the form of charts are displayed at the treatment plant for easy access by the maintenance team.

Breakdown Maintenance: These are maintenance activities carried out as a result of a breakdown or failure of some part. If the breakdown requires a minor component repair it is considered under Operating Expense and if the breakdown requires a major component repair it is considered under Capital Maintenance Expenditure.

Financing of Capital Maintenance Works

Generally, financing capital maintenance works is a huge challenge for most water utility companies. This is because there is no working reserve fund to carry out these works. Some utility companies are usually able to generate funds internally through revenue collected from customers. However, less revenue is usually obtained than it is expected. Since revenue generated is low, it is unable to cater for all the operation and maintenance activities of water utility company and there is no reserve fund to finance all capital maintenance works. Governments should therefore give some allocation for water and this should be used for system extension and some maintenance works. Asset maintenance, therefore, is a huge burden on water service provider. Some challenges may be experienced in performing the large capital maintenance works. This is due to inadequate funding from internally generated funds and donors. Huge capital maintenance works are financed primarily through concessionary loans. The process involved in securing funding through concessionary loans is lengthy. Work requiring huge capital investment is funded by loans and grants obtained from foreign bodies secured by responsible governmental boards.

Factors Influencing Capital Maintenance Expenditure

Ideally, it is appropriate to identify the factors influencing capital maintenance expenditure both qualitatively and quantitatively. These factors may be identified from two perspectives namely: Factors that influence the magnitude of expenditure and factors that influence the magnitude and frequency of capital maintenance works. Some of these factors are outlined below.

Choice of Credit Facility: Two options are available for financing capital maintenance works in most countries around the world. They are cash and debt financing. Due to unavailability of reserve funds for capital intensive maintenance, most utility companies usually resort to debt financing for some capital maintenance work. Interest charged on the loan influence the magnitude of capital maintenance expenditure. The government should pay more for the works carried out than it would have paid if it had financed the works through cash financing.

Use of Foreign Expertise: Some capital maintenance works require a high level of technical expertise and the use of complex equipment and technologies. Where the technicality and magnitude of works is high, foreign companies are used in performing capital maintenance works in the water sector because they are more experienced in carrying out huge maintenance works. These companies charge rates in their national currency and for their service. Since these personnel are expatriates, their charges include accommodation and feeding and this increases the total contract price.

Size of Water Supply System: Water supply systems with higher design capacities usually have a higher magnitude of capital maintenance expenditure. The size of water supply utility company also influences the magnitude of capital maintenance expenditure.

Age of Water Supply System: As the components of the system ages, the asset and their components wear out and must be repaired or replaced. Very old systems are less robust and usually need more repairs and replacements. Sometimes, their parts are obsolete and thus it is necessary to replace them with parts readily available on the market. This may require changing of certain components to accommodate for the new replacement.

Delay in Carrying Capital Maintenance: Studies have revealed that delaying repairs and replacements influences the cost in two ways; (1) postponed repairs and maintenance accumulates in number and magnitude and contributes to a more complex problem (2) cost of maintenance changes due to the time value of money. Reasons for delay in executing capital maintenance work at water supply systems may include lengthy tendering procedure, unavailability of funds and unavailability of spare parts. Since funds requirements for rehabilitation and expansion of work that is needed to be carried out at water supply systems are not readily available, most of the works are usually pending and their cost kept on increasing with time.

Erratic Power Supply/ Power Surges and Outages: Most water supply systems use power from the National Grid as the only source of energy. Increased number of failure of some equipment and parts of water supply systems may be experienced especially because of frequent power outages and power surges. Usually, when the power supply surges, is erratic or failing, it causes the breakdown of certain equipment and thus the need for their repair and replacement increases.

6 AN EVALUATION OF WOMEN ROLE IN WATER SUPPLY SYSTEMS

INTRODUCTION

Water is an indispensable resource not only for the sustenance of all life but also to human development; people rely on it to grow food, produce goods and generate energy. Furthermore, water is a vital element for healthy ecosystems affecting biodiversity, livelihoods, health and education. There is simply no substitute for water as every living thing depends on it. Given its importance, it is the responsibility of every citizen to try and manage as well as preserve and protect water resources. However, non-participatory management and development of water resources remains a hindrance to sustainable human development, growth and poverty reduction. It has been suggested that the present water insecurity in different countries around the world have neither resulted from natural limitations of water supply systems nor is it from the lack of appropriate technologies and financing but rather from reflective failures in management and preservation of water resources.

Source: http://www.rwsspwn.org.np/cross-cutting-gesi

Figure 6.1: Gender and Social Inclusion in domestic water supply

Different regions have their own vision towards have water security. For instance, African Water Vision (2025) focuses on "*equitable and sustainable use and management of water resources for poverty alleviation, socio-economic development, regional cooperation, and the environment.*" The vision further recommends gender mainstreaming in water resources management with a special call on women to occupy key positions and functions in decision-making processes on water issues.

Women's involvement and participation in water implementation programs; especially in community level is largely viewed by many people as key players in promoting sustainable domestic water supply. For instance, in some communities, women are usually responsible for the construction of water storage structures such as sand dams.

Source: http://www.indiawaterportal.org/articles/gender-inclusive-development-mountains

Figure 6.2: Women taking part in water related development projects at community level

Usually, this type of community work is usually based on voluntary services which are based on the women's "free time". Generally, the women's work and contributions still lack the recognition they deserve. Whenever water management enters the public domain, the men take up the leadership positions and take the lead in decision making processes in the water committees. As a result, the women are excluded from management activities. Women's exclusion from decision making often leads to the development of water projects and/or water programs that are nonresponsive to women's practical and strategic needs, and reinforces gender inequalities in the rural communities. Women are not given an equal voice in decision-making on some key issues which directly affect their lives, especially those related to water resource management. The solution to these problems therefore lies in breaking the cultural barriers so as to ensure that women are equally involved not only in the project identification processes, but also in the actual planning, design, construction of water supply structures and in the overall water resource management activities. Therefore, local gender contexts ought to be considered when developing the water development policies since each situation is unique.

WOMEN'S ROLES AND RESPONSIBILITIES IN WATER RESOURCES MANAGEMENT

Studies have shown that most women are rarely consulted in project identification and formulation planning which are responsibilities of men. However, it has been argued that the types of projects which are needed to be undertaken in a community need to be based on the recommendations of people who live most of their lives in the areas concerned.

Women should be involved in identifying the important water projects needed in the various communities. According to WHO (2010), most women usually have high sense of knowledge in all aspects of the living conditions of the people including what is conventionally termed "water engineering" topics which are traditionally associated with formally trained engineers. Contrary to this western concept of engineering, women are very knowledgeable in the local climatic, hydrological and drainage conditions.

Source: https://www.mcc.gov/blog/entry/blog-071615-conserving-jordans-critical-water-supply

Figure 6.3: A group of women taking part in a water conservation campaign meeting

Women are rarely informed about important meetings where decisions on water project planning and implementation is discussed. Such meetings are announced in networks comprising only of men who draw up the plans with dates of implementation which are never discussed with the women. The reason is that such projects are considered as the territories of men only. Thus, any issues on tendering of contracts in the water projects only circulate among the male population in the community.

Women participation in the implementation and management of water supply system can be very vital in provision of basic information in the strategic decisions. Such strategic decisions include choosing new technologies and manpower. In terms of monitoring the impacts of water projects in community, women can be used to report cases of broken down equipment, water spillage, poor workmanship and other problems to the water project managers who should then take the strategic decisions for addressing the problems. There are several challenges that hinder women participation in the water supply system in a community. These challenges are discussed below.

Socio-Cultural Challenges: Different studies have shown that traditional practices and norms are usually key hindrance for women's participatory role in making decisions that influence implementation and management of water supply systems in the community level. Up to today the power differentials within the community generally dictate who participates in various meetings. For instance, it has been seen regularly that men, who are commonly community leaders, are invited to participate in consultation meetings to discuss the design and formulation of new water projects. Therefore, men are found to have the exclusive power in making key decisions on behalf of the entire community without any input from the women of whom however, represent the perspective of water interests.

Therefore, it is important to put in mind that some community, especially in developing countries, usually believe that it is socially inappropriate for women to actively participate in public spaces. It also needs to be mentioned that it is also considered unfeminine behaviour for women to challenge men's decisions at public forums. Consequently, women are discouraged from speaking in public forum. Another cultural impediment to the participation of the women in public meetings is that the local men tend to have some sort of entrenched territorial claims over the women consequently curtailing the women's freedom and mobility. This makes it difficult for most women to get the opportunity to attend water project meetings and training beyond their own confined spaces, that is, the homesteads. Therefore, because most of them usually do not have viable means of transportation, most of them are usually limited to their abilities to actually have access to important meetings where crucial activities involving water supply and management systems are discussed.

***Economic Challenges**:* It has been indicated that poverty among the women households also constitutes a constraint on their participatory role in community water projects. Economic power, in terms of income levels, is a key factor that influences the weight the communities put on the suggestions coming from people regarding public projects. Since the bulk of the women usually operate in the subsistence sector, they are generally low income earners. They are therefore not considered as the bread winners of their families.

***Women's Confidence**:* It is evident from several studies that there is usually a combination of factors influencing the extent to which women feel motivated to participate in water management. These factors include: limited skills and experience and lack of confidence. It has also been shown that general women's lack of education presents a formidable barrier to their participation in the water management projects. Most women in the community usually have relatively higher illiteracy levels which apparently prevent them from occupying important leadership positions in the communities in comparison to their male counterparts. Studies have also revealed that the illiterate women in community generally lack confidence in their own abilities because they are unable to read and write. In addition, women's lack of confidence is usually observed during meetings where all the men usually sit at the front benches whereas the women are usually seated at the back.

RELEVANT THEORIES AND LAWS THAT EXPLAIN WOMEN'S PARTICIPATORY ROLE IN WATER SUPPLY SYSTEM

There are numerous materials in the literature discussing on the rights of women to access clean, safe and adequate water supply in their households. However, there is little information on how these rights can be incorporated and recognised as part of socio-economic rights in the community. Most of these rights are mainly focussed on women's rights to access water, in what is considered basic human right. This access to water is primarily for domestic uses such as cooking and washing and water for small scale and substance agricultural production. Therefore, women act as key players in supplying domestic water in households. There are several theories that can be put into practise in explaining women participation in water supply and management practices. According to Marxist feminism theory, most women are usually not free economically and they exist in a capitalist society which combines with male dominance to subjugate most women in the community. In other words, women are usually oppressed when a community is divided based on capitalism. This relegates the women to the non-economic private sphere.

This issue can actually be changed through altering the social-economic systems in the community or country in order for the women not to be marginalized, dependent, and exploited. Hence, women face challenges in breaking into other work not associated with domestic work like work in the informal sector. The conditions at the markets contribute to how women's work is perceived and undervalued. It is also important to analyse liberal feminism and Ecofeminism because some of the assumptions on women participatory role in water supply system usually touch upon how gender neutral laws result in the undervaluation of women's work.

Liberal feminist states that "*subordination of women is set in legal constraints that block women from fully achieving their potential in the public world*." Women have special needs and are uniquely different from men. If their differences continue to be ignored by the legislature, women will continue to be subordinated in society. In most countries around the world, water legislations usually fail to acknowledge this link of women subordinated roles and gender neutral laws. Liberal feminists insist that "the *solution to the oppression of women is to remove all formal barriers to their equal participation in social, political and economic life, thus providing equal opportunity for all.*"

Thus, the law can be used to improve the situation of women in access to water in a place if it acknowledges the gender dimensions involved in accessing resources. If these gender dimensions are ignored there are several consequences which go towards the undervaluation of women's work or limits women's work capabilities. According to WHO (2010), this is well put as follows "*By overlooking social and gender differentiation in relation to allocation and uses of water, the seemingly neutral laws and policies may easily be to the effect that existing social and gender inequalities are reinforced and increased.*"

The other theory is ecofeminism. This explains the interrelatedness of women's oppression and that of nature. According to scholars, the word ecofeminism "*was first coined by the French writer Francoise d'Eaubonne in 1974 to represent women's potential for bringing about an ecological revolution to ensure human survival on the planet.*" This theory can actually be used to explain how women in the society handle the effects of insufficient water at their homes.

Ecofeminists are usually concerned with enhancing the relationship between the human beings and the nature bearing in mind that both the nature and women usually have a mutual association and they can actually be devalued when they are both liberated into a political situation. The male designed development models neglect the effects of insufficient water to environmental protection and on women's capabilities in development and their working lives. It has been stated severally that "*Ecofeminism has philosophical, political activism and spirituality aspects. In other words, Ecofeminists can use all these to question the male bias in environmental movements and development models and campaign for animal rights and against reproductive technologies.*"

APPROACHES IN WATER SUPPLY SYSTEMS

There are several approaches available for water supply system, especially in the rural areas. These include conventional communal water supply sources, self-water supply sources, and also domestic water supply sources.

Conventional Communal Water Supply Sources: In most countries, the conventional communal water supply sources are usually considered as key water points within communities. However, they receive very little or no funding at all in order to achieve the Millennium Development

Goals in water targets of a country or community. Nevertheless, it worth noting that coverage of water facilities has increased in most countries around the world due to NGO's initiatives in water and sanitation programs, some of these facilities have been abandoned by beneficiary communities due to poor governance. This poses a challenge to most people considering the fact that the majority of population live in rural areas. As a result, most people consider self-supply initiatives as the best alternative approach to water supply construction and management in rural communities.

Source: https://blog.usaid.gov/category/water-and-sanitation/

Figure 6.4: NGO's initiatives in water and sanitation programs

***Self-Supply**:* Self-supply builds on the initiative of individual households or communities to improve domestic water supply through user investment in water treatment, source construction upgrading and management. This is usually based on locally available and easily affordable technologies to the users in rural communities. Self-supply initiatives are spearheaded by beneficiary communities who have the income and are willing to contribute towards the construction of domestic water supply sources. Since most of the people who live in rural communities are poor, they sometimes mobilize their neighbouring communities to pull their resources together using local labour, materials and initiatives towards the construction of domestic water supply sources.

It is worth noting that such sources are often associated with poor water quality and seasonal unreliability. It is also worth noting that although the self-supply initiatives are private, access and use of the domestic water source by other households is usually at no cost or for a small fee as a way of promoting social relations. This is because water is seen as a natural resource and as a result the payment of a water fee in the rural areas are not acceptable. This system can compromise access of water among the disadvantaged groups in society especially women who do not have the capacity and ability to construct and or maintain the domestic rural water supply sources.

Domestic Water Sources: According to UNDP (2006), there are five patterns of water use among households in most places around the world. For instance, in most rural villages, most of the poor households usually draw water from a protected well for part of the year but then they are forced to draw water from rivers or streams during the dry season. Studies have shown that water drawn and used in a day depend on factors such as availability and perceptions of quality. Additionally, it has also been noted that the use of water sources varies temporarily and seasonally, due to changes in water quality and pressure. Example are some rural communities of Indonesia where groundwater cannot be used for drinking because of salinity and pollution. The UNDP (2006) has suggested that in such circumstances groundwater should be used only for cleaning or washing. What imaged from a large group of countries is that patterns of water use are far more complex and dynamic than the static picture presented in global reporting systems. The real life patterns constantly adjust to consider concerns of water quality, proximity and reliability. It has also been noted that in most poor rural households, people get their water from a variety of sources such as lakes, streams, springs, wells and rivers. Protected villages wells are the most common water sources.

PROBLEMS THAT WOMEN FACE IN SUPPLYING WATER IN THEIR HOUSEHOLD

There are several problems that women face in supplying water in their households. One of these problems is the issue of the time they spend collecting water from the source point. This problem usually represents the heaviest burden on most women in the community. For instance, in countries such as Uganda, Kenya, and Senegal, women usually spend on average of 20-35 hours per week fetching water for domestic purposes. According to research study that was conducted in 2006 by United Nations Development Programme (UNDP) that in a year, about 40 billion hours are usually spend by women in the Sub-Saharan Africa collecting water from different sources. This actually represents one year's labour force for the entire workforce in France. Reducing time for other activities such as child care, rest or reproductive work, the time spend collecting water affects family and self-time efficiency, disempowers women and lowers their income base.

Source: http://www.ipsnews.net/2012/08/women-spend-40-billion-hours-collecting-water/

Figure 6.5: Women spend more than 40 billion hours annually in Sub-Saharan Africa collecting water

It has also been noted that the total of women's working hours for collecting water are usually fewer in communities with water provision as compared to those living closer to the water source. Hence, having access to water infrastructure can reduce the time burden on women. Researchers have actually noted that this is not implicit; however, the time women save on water collection would be devoted to paid for other activities. It has been suggested that additional policies are needed to achieve that goal, especially policies related to building human capital and provision of child care facilities.

Another key problem is distance to water sources. According to the United Nations Population Fund (2002), in some regions of the world, such as parts of Asia and Africa, women have to walk great distances to find drinking water. In some African communities, a hand pump 500 meters from the household is usually considered as a luxury. Studies show that by the year 2000 it was estimated that one-sixth of humanity (1.1 billion people) will lack access to an improved water supply within 1 kilometre of their home. Both women and girls are usually responsible for this task. On average, the distance a woman in the developing countries walk to get water is about 4 miles a day carrying a load weighing 40 lbs. This is a long way to go and a lot of water to haul. According to WHO (2010), about one third of women in Egypt walk more than an hour a day to collect water, in other parts of Africa, women spend as much as eight hours collecting water. It is further noted that the average distance walked by women in Africa in search of domestic water is about 6 kilometres a day. In Asia, an Indian national survey for UNICEF found that women spent an average of 2.2 hours per day collecting water from rural wells far away their households. In a study done in Sri Lanka, which is considered to be well provided with water sources, found out that 10% of women had to travel more than 1 kilometre to their nearest water source. In a study on water supply and sanitation in Gouansolo village in rural Mali, 65 kilometres South of Bamako the national capital, it was noted that average distance travelled to water sources is 44 metres, the closest is 3 metres and the furthest is 260 metres.

Queuing time to collect water is another major problem that most women are faced with. Studies have indicated that excessive queuing times are indicators of insufficient water availability, either due to an inadequate number of water points or inadequate yield of water points. Long queuing time also has potential negative results of reduced per capita water consumption, increased consumption from unprotected surface sources, and reduced time for water collectors to tend to other essential survival tasks. A research done by WHO and UNICEF (2010) revealed that those spending more than half an hour per round trip progressively collect less water, and eventually fail to meet their families' minimum daily drinking water needs.

Source: http://www.theweek.in/content/archival/news/india/Woman-dies-waiting-in-queue-for-hours-to-fetch-water-in-Latur.html

Figure 6.6: Long waiting time in water queues

Another problem that women face in supplying water in their households is the effect of multiple users. The driving principle that domestic water can be used for livelihoods refers to take peoples multiple water uses. In other words, instead of using a single source for end use, multiple use of domestic water is used in most places as a way of achieving the benefits of multi-faceted livelihoods.

CONCLUSION AND RECOMMENDATIONS

Studies on water supply systems have revealed to have implications on women in rural households. According to World Bank (2005), there are gender inequalities in domestic water provision. Improvement in access to water can only be possible if women are empowered to participate in decision-making processes at the community level. This is particularly in aspects such as household water supply, provision, water use, management and maintenance. Special attention should be given to the active involvement of women by giving priority to their needs and ideas in the planning, implementation and management of domestic water projects in these two communities.

Studies have shown that women are the principal water collectors and managers of domestic water at household levels. Nevertheless, it is worth noting that in most instances they are faced with different obstacles such as lack of financial resources, technical capacity and managerial expertise required to ensure that water sources in their area are fully functional. As a result, they are usually affected negatively throughout the domestic water chain. Therefore, there is a need to investigate the water yield at each water source to determine the number of households that each water source can support. Major emphasis should be on water quality and quantity as well and the possibility of using traditional water treatment methods such as the use of ash in place of aqua tablets in rural communities.

The development and implementation of the domestic water supply projects should be based on the needs of women and participation of all groups including women who serve as principal water collectors and managers at household level. The community should elect an active water user committee whose role should be to mobilize and coordinate the communities on the planning management and maintenance of water sources. Investment in self-supply should be given priority as it is bound to bring water sources near households thereby helping in reducing the distance that women in the two communities travel to access domestic water. It is anticipated that this will give women the opportunity to participate in other productive activities. However, most of the women in most places around the world are usually poor and therefore there is need for them to be empowered economically through the use of domestic water supply sources.

It is important for government and donor agencies to focus aid on promoting self-supply water projects at household level which are made convenient and effective. This should be a much more inclusive approach that benefits women in community. It should also be in a manner that does not limit women's social networks among themselves. Women as principal beneficiaries should be willing to own, operate, manage and actively participate in government and or NGO/CBO initiated projects on domestic water supply. Such projects should include financial resource mobilization, water systems management and operation that should be aimed at empowering women in water service provision and management in community.

Table 6.1: Some of the recommendation towards enhancing women participation in water supply systems

Recommendations	**Key Points**	**Expected Outcome**
Gender Planning	• Redefining the role of women in the community • Defining the role of women in the community water management • "women as real partners with men" • Gender-specific policy formulation on women's role in community water management and distribution	• A more direct approach to the everyday water management problem of the community • Women empowerment and participation in the improvement of water management in the community • A more practical approach in water distribution
Increasing water sources	• Inclusion of a definite increase in water sources in as early as the planning stage of a community water program • Increase water supply for the sole purpose of providing benefit for women • Consideration of the multiple water use service approach to designing water sources	• Positive influence on livelihood and domestic activities of women, household income and househols spending pattern • Reduction of households per water source • Reduces time spent for queuing for water ot travel to its source • Reduces competition among neighbouring communities
Conducting Researches	• Analytical determination of the link between domestic water supply and household expenditure patterns • Determination of the extend of remittance influence in rural economy and extend of its impact	• Educating women on how to be more objective in looking into domestic expenditure patterns • Give space to women and open them to a more technical and systematic way of solving water-related livelihood activities

Water Policies	• Decentralization of management and implementation of policies in the community level • government should have one single authority or a department that is responsible for the coordination and development of domestic water provision at grassroots level. • government should consider having two policies dealing with domestic water supply. One policy should be focusing on domestic water provision in rural communities and the other should be focusing on domestic water provision in urban areas	• Decentralization of policy would ensure equitable distribution of water resources for the benefit of recipient groups in different communities. • a strategy of having two policies catering for two different segments of the community will ensure an equable distribution of resources and development.
Social Work and Women Participation	• social workers be allowed to work with women particularly with their socio-economic and social problems • empower women by showing them how other women contribute to the society	• social workers can go further to analyze the cultural and traditional aspects that predispose women to their current situations which are part of women's experiences.
Community Social Relations and the Social Worker	• establishment of social relations and social networks is important for peaceful co-existence in rural communities • Social workers should take a lead in educating rural communities about gender equality and gender equity	• increased demand of the water resources which may lead to conflicts among and between communities • Social workers would then be expected to take steps to facilitate dialogue among or between conflicting communities, to facilitate positive interaction

Source: http://slideplayer.com/slide/6377070/

7 PRIVATIZATION OF WATER SUPPLY SERVICES

INTRODUCTION

The significance of water in life cannot be overemphasized. According to studies, more than one billion people across the world, especially in various developing regions have no access to clean and safe water despite the initiatives taken by the United Nations to emphasise that water access is a basic human right. According to World Bank (2000), low income households have a limited access to water and pay more compared to middle and high-income households. Since water is a public commodity, it is the responsibility of the government to ensure that the public has the access to safe and clean water through the enactment and implementation of appropriate public policies.

Reforms, Local Government Act and Water Acts are some of the key legal statues that should be put in place in order to help in governing water sub sector in a country. In most of the countries, local authorities are usually empowered and acts as the representative of central government to acquire, establish and maintain supply of water services within their area of jurisdiction.

However, in some countries, local authorities are faced with different challenges in executing their service delivery. Sometimes there are issues of misappropriation of funds collected from water service delivered, mismanagement, and inefficiency in their water service provision. In some developing countries such as Africa and some parts of Asia, water supply services are characterized with frequent water disruptions, billing problems, poor water quality and unaffordability concerns for the urban poor. To reduce these problems in water service delivery, there should be broad framework of water sector reforms that should be based on a philosophy of separation of roles and responsibilities in terms of policy, regulation and service delivery which should represent a new institutional dispensation. This act should be implemented alongside privatization of water services in various local authorities in a country. Privatization of water services in a country should be done through the formation of a company that operate under the guidance of market principles. This is in accordance with neo-liberal dogma to treat water as a tradeable commodity as opposed to a public good that government should provide to its citizens. According to the World Bank (2005), transformation of water services to a tradeable commodity means that, access to water is now pegged on consumer ability to pay as opposed to need.

To examine the efficiency and effectiveness of privatization process, consumers' access to water services is one of the ideal evaluation criteria. This chapter is going to assess privatization of water services and its impact on access to clean and safe water.

PRIVATIZATION SCOPE ON WATER SERVICE DELIVERY

Privatization process is informed by the desire to achieve efficiency, effectiveness in provision of goods or services. According WHO (2010), privatization of water service delivery usually helps to strengthen free market economy through state withdrawal and enabling multiple private actors to venture into market for the competition to thrive. This would culminate to reduction in price of goods and services inevitably offering consumers with more options at a reduced price. In water service provision, efficiency and effectiveness is achieved through cost recovery measures that entail increasing water company's ratio of revenue collection against operational and maintenance costs.

To ensure efficiency, water companies resort to cost cutting measures by introducing water fee where there was none, increasing water tariff and investing in areas that promise profitable returns. When efficiency measures are implemented, access to water services by low income households is compromised. In other words, reduced water service access can be attributed to the unwillingness of the private water companies' to serve those households in the low-income class who are usually challenged economically to pay for the water services offered to them.

The introduction of water charges means that access to water services is determined by consumers' ability to pay as opposed to needs. However, water is a basic human need (right) whose access is paramount irrespective of one's ability to pay. But, there is very little literature on privatization of water services delivery and how it impacts low income households with respect to water services access.

Studies reveal that private water companies have the tendency to increase price of water and only invest in profitable high-income areas. Failure by the private water companies to expand and rehabilitate water network in low income areas raises concerns if, whether low income households can still access water services with onset of privatization. According to World Bank (2010) reports, provision, delivery and access to water services can help in improving major local authorities that have implemented the privatization of water services.

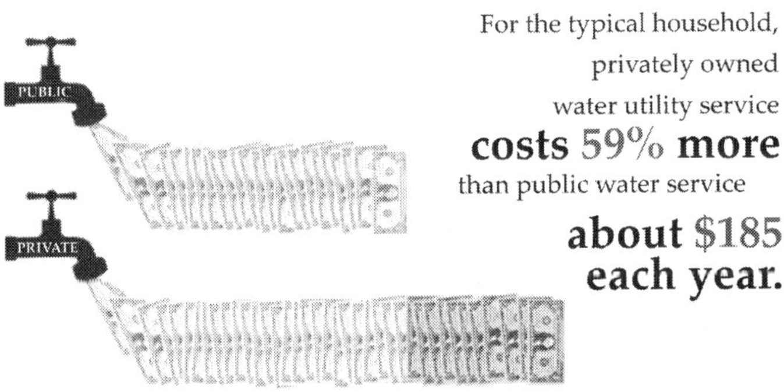

Source: http://ianchadwick.com/blog/council-is-privatizing-our-utilities/

Figure 7.1: A typical comparison between private and public water companies based on water costs

The successes and efficiencies of private water companies is stipulated on the efficiency of revenue collection, reduction of unaccounted water, improved billing. Lack of information on how privatization of water services impact on access to water services by low income households usually lead to a knowledge gap. Acknowledging the market measures introduced with privatisation of water services, and access to water services by low income households are one of the main problems. There are certain questions that must be asked when discussing the issue of privatization of water service delivery. These questions include how does privatization of water services impact access to water by low income households? How does privatization of water services affect the availability of water services for low income households? And how does privatization of water affect cost of water for low income households?

MECHANISMS FOR PRIVATIZATION OF WATER SERVICES

There is no universal definition of privatization scenario which has contributed to a myriad of understanding of the process both in theory and in practice. Studies provide a detailed account of mechanisms through which privatization process can take shape. It has been argued that privatization process should be understood in two different ways based on narrow and broader definitions. In a narrow definition, privatization process mean sale of government owned company's assets or shares to an individual/s or private firm/s with purpose of generating capital. In a broader meaning, privatization is conceptualized to mean the introduction of policies that increases the role of market forces in the provision of goods and services. To permit market forces in determining distribution of goods and services may entail the following mechanisms; corporatization, lease contract, service contracts, management contracts, concession contracts, public private partnership, built own transfer-contract, franchise, deregulation and control, introduction of user charge, grant system, voucher system among others.

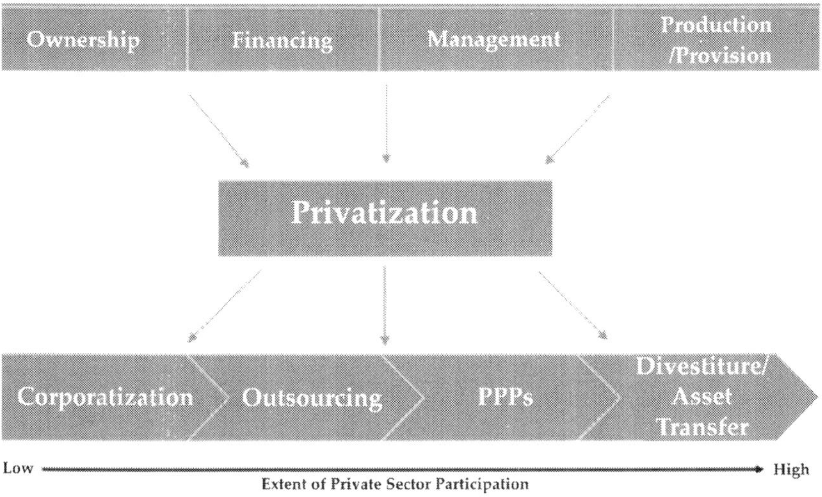

Source: https://www.researchgate.net/publication/313089612_What_is_privatization_A_political_economy_framework

Figure 7.2: Processes of privatization of water supply systems

Corporatization is a mechanism that entails government establishing an independently managed company run on commercial principles in the provision of goods and services. This sees government agencies transformed into market players competing with other players in the industry in the provision of goods and services. For instance, corporatization of water services has been employed in Nairobi Kenya as one privatization model. Corporatization of Nairobi water service led to an increase in the cost of water for low income households as a result of upward water tariff review. The water company did not invest in low income residential areas since they could not guarantee profitable returns. Investing in such areas could be a setback in realizing full cost recovery which is the major motivator to privatization of water services. They further contend that privatization through corporatization is a neo-liberal strategy to make government invisible for public accountability.

Lease contract in privatization of water services has been employed in Guinea. This mechanism entailed inviting a partnership between Guinea government through a public holding company (SONEG) as minority shareholder and private company (SEEG). In the lease agreement

SONEG owned water assets was in charge of investment planning while SEEG was responsible for maintaining and operating provision of water services. Major improvements were recorded with regards to improved water quality and connections, improved billings, increased metering to Conakry residents among other efficiency measures. However, there was little information on how access to water services by low income households of Conakry Guinea was impacted with advent of privatization of water service. Studies have also shown how privatization of water services was met with opposition from a coalition of trade unions and human rights activists in Ghana. Decision makers had failed to consult citizens and other stakeholders in securing a lease contract in privatizing Ghana's water services. After the negotiation on the best ways to solve perennial problems in water supply in Ghana was over, management contract was finally adopted to improve provision of water services. A public agency was tasked with the responsibility of asset holding and investment in the sector while an international private entity was tasked to operate water supply system. Regardless of the mechanism adopted in privatization of water services, improved efficiency and access to water services is the ultimate goal. Introduction of market and commercial principles in provision of water services is evident as observed in different mechanisms adopted in the privatization of water services.

HOW PRIVATIZATION AFFECTS ACCESS TO WATER SERVICES

Introduction of commercial principles in the provision of goods and services in effect transforms a basic need which ought to be provided for free, or at subsidized rate, to a commodity offered based on consumer ability to pay. Different scholars have examined privatization of water supply. Through privatization of water services, most water companies usually increase water fee across consumer type. For instance, low income households are usually not able to meet the increased water tariffs. The inability of low income households to pay the increased water charges leads to disconnection of water supply to their houses thereby denying them access. To safeguard low income households, the national government in different parts of the world usually introduce a subsidy program and install alternative water metering for low income households to safeguard access to water.

There is need to examine the reasons why the public's options for water service provision usually perform dismally to warrant the introduction of market forces through privatization. For instance, in a country like Rwanda, privatization of water services has been implemented as a policy prescription of structural adjustment program. Privatization of water services does not guarantee efficiency and improved access to water services if not augmented with capital investment. According to World Bank (2010), prolonged economic underperformance and reduced capital investment in water sector is what hampers local authorities' mandate in water provision. In other words, inadequate capital usually compromise local authorities to expand and rehabilitate ailing water network. In most cases, privatization is thought to have impact of injecting the much needed funds to overcome inefficiencies in water sub-sector.

For instance, in a country like Ethiopia, more than ten private water companies were established as a remedial measure. However, privatization did not solve the problems witnessed in the public mode of water provision due to lack of requisite capital to invest in water infrastructure. Whether water service is a public or private responsibility, heavy investment in the sector is a requisite. According to World Bank (2010), financial capital is mandatory for an improved water service provision.

Source: https://cityvoiceng.com/nlc-groups-protest-against-privatization-of-water-in-lagos/

Figure 7.3: Illustrating protesters in Ghana against privatization of water supply systems

Privatization of water services is usually not embraced without opposition from some quarters. Opposition to privatization of water services is based on human rights claim that water is a public good that cannot be bestowed on private hands for its supply and delivery. For instance, in Ghana, privatization of water services was met with civil disobedience of angry protesters in the street over increased cost of water. Increased water fee had disfranchised low income households access to water. Ghanaian government had to prematurely terminate the contract entered with multinational private water companies for provision of water services in the city. This is yet another case of failed privatization. While Ghana's case illustrates a failed privatization that was reversed, this chapter examines an on-going process with special emphasis on access to water, especially by the low income households.

Successful privatization of water services requires one to factor special needs that characterize a system targeted for privatization. This scenario has been experienced in Yemen which led to an improved access to water services by all kinds of consumer type. In this country, water management was decentralized, autonomous water corporations established, a clear service regulation set up and private partners enabled to participate in the sector. The reformed measures in water sector increased level of access to water by 90% of the population.

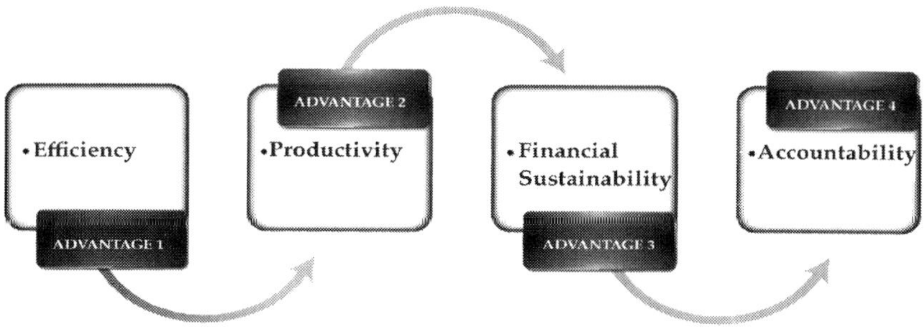

Source: http://www.wasa.gov.tt/WASA_AboutUs_CorporateGovernance-Background.html

Figure 7.4: Illustrating some of the major advantages of privatization of water services

According to the World Bank (2010) report, it has been established that privatization of water services increases access to water services by high income households than the low income households. Inequality in the access to water services by low income households is, because of private water companies' failure to expand, extend or rehabilitate the water network to low income residential areas. This inequality is also observed on the type of water service providers for different categories of households. For instance, it has been noted that middle and high income households usually depend on corporate organization for their water needs while low income households depend on water vendors. However, water vendors are usually more expensive than households with inn-house water connection.

In another study conducted by UNICEF (2011) on households' water use and unit cost of water in Kenya's three major cities of Nairobi, Mombasa and Kisumu, it was found that both low income and high income households do not have adequate water supply. In these three cities, it was established that primary sources of water were mainly kiosk, yard tap, water vendor and private water connection. It has been revealed in other previous research studies that water kiosk and water vendors usually charge more than the conventional in-house connection. Additionally, most low income households usually fetch less and spend more time sourcing for water compared to middle and high income households. It is essential to also note that both low income and high income households in some places, especially in developing countries are usually dissatisfied with quality of water supply in general. Most people usually express the desire for an improved services provision from their main sources of water.

There are cases where low income households usually have improved access to water after privatization process. In a previous research conducted in Uganda, it was revealed that privatization process did not deny low income households access to water. Improved access by the low income households was made possible with introduction of subsidy programs and with the extension of water network to underserved and unserved low income residential areas. It has been argued that extension of water network and facilitating water connection in slums reduces cases of illegal water connections.

At a household level survey conducted in Thailand, it was revealed that there's an increased access to piped water services by low income household after the privatization of water services. In the study, data collection procedure was tailored to find out changes in level of access to water prior and post privatization era targeting high, middle and low income household respondents.

Provision of a utility service like water through public or private mode is a policy option pursued by governments. In a research conducted in South Africa's Cape Town, it was revealed that apartheid legacy of territorial discrimination along racial lines continued to determine local authorities' utility service provision. Classification of residential areas based on inhabitants' race had that, the upper strata mostly white, enjoy a better water service provision during the pre and post privatization of water services.

After a period of water reform that involved adoption of market principles in water service provision, Cape Town local authority corporatized water sector. However, established water company inherited the apartheid era water network system that favoured high income households. This in effect denied access to water services for black people (whose majority were low income households). Privatization of water services did not change much since water network favoured high income households at the expense of low income households.

Due to unreliability of piped water system or lack of it, water vending has become a prevalent feature as a source of water for households. The practice thrives even after it has been established that it is more expensive than piped water system. According to World Bank (2005), households who rely on water vendors spent more than 10% of monthly income on water, more than the recommended 5% of households' income.

Different scholars have actually demonstrated how a market reform through privatization does not guarantee the ideal free market economy in water sector. It has been demonstrated that market intervention is not the appropriate remedy for a monopoly good like water supply. Inefficiency of water municipal council department in different country should be used as a prime justification for privatization of water and waste water services. However, privatization does not solve the inefficiency problems that are witnessed in municipal council water department. Instead the private water companies are usually characterized with operational inefficiencies, lying off workers to reduce costs, numerous boardroom wrangles and other institutional failures. The inefficiencies that are witnessed in private water company are the reasons most municipal councils in different countries around the world usually de-privatized their water service, especially when the private contract expired. Privatization is therefore not the sole solution to myriad of problems witnessed in water sector. It is important to note that market is never perfect and there are cases when market fails to deliver.

A research conducted in Kisumu County in Kenya by World Health Organization revealed that water vending is a prevalent feature in low income residential areas due to unreliability of piped water system. Lack of resources to expand water infrastructure to serve low income households also explains the existence of water vendors as alternative water service providers. While this study was more concerned with privatization and access to water, it equally looked at the role of water vendors as source of water for low income households in the city.

Based on what has been discussed above on the impact of privatization of water services delivery, it can be concluded that increased access to water services through privatisation of water services delivery, especially by low income households can be enhanced through multiple of intervention mechanisms. These interventions mechanism include; introduction of affordable water tariffs, introduction of subsidy programs targeting low income households and investment and expansion of water network in un-served and underserved areas. On the other hand, reduced access to water services by low income households is attributed to increase of water charges and failure by water companies to expand water network to low income residential areas.

ROLE OF REFORMS IN PRIVATIZATION OF WATER SUPPLY SERVICES

Water supply services are usually governed by the arrangement of different governmental institutions. Different countries have different institutions that are mandated with various responsibilities of managing and sustaining water services. For instance, in a country like Kenya, there are six key institutions mandated with these responsibilities. These include the Ministry of Water, Water Apportionment Board, Catchment

Boards, Water Resources Authority, Local Authorities, Regional Water Committee, and the Water Undertakers.

The Ministry for water is usually mandated to investigate, conserve and promote proper use of water resource and control over water upon any land in the country. On the other hand, Water Resources Authority is responsible of advising the water minister on establishment of catchment areas and appointment of Catchment Boards across the country. The Apportionment Board usually has the responsibility of issuing water permits to various water undertakers.

There are major challenges the entire framework of reform in privatization of water services is facing. For instance, the powers to appoint members of various water institutions are usually associated with too much nepotism and patronage. It is also worth noting that lack of specific criteria in the appointment of various offices usually compromise professional service delivery which in effect would cripple performances of these institutions. There are numerous institutions that can be merged together in order to have an effective management of water sector. For example, in Kenya, the Water Resources Authority and Regional Water Committees both usually have the responsibility of advising the minister on water conservation.

Provision of water services in most countries around the world is usually subsidized by the central government. The existing water tariffs are usually too low thus undermining sustainability of water supply. Introduction of market reforms under the auspices of structural adjustment programs can help central government to reduce financing the water sector. According to World Bank (2005), reduced funding to water sector coupled with multiple devolved functions and poor management practices may contribute to local authorities offering inefficient services. These challenges may cause various local authorities to experience unreliable water supply through frequent water disruptions, poor water quality, billing problems and high costs of water. These deficiencies are what contributes to poor performances by public sector in delivery of water and sanitation services in a country. Multiple institutions that lack clear roles separation, centralization in the management of water sector and poor service delivery are usually the main reasons to necessitate reform agenda in the privatization of water services delivery.

EFFICIENCY IN WATER SERVICES DELIVERY PRIVATIZATION

When evaluating the efficiency in privatization of water services delivery, it is important to look at various independent sources on performance of water companies. Privatization of water services is a progressive process that may lead to the formation of different water companies. Water companies that have been privatized should operate under the guidance of various market principles. For instance, private water companies are supposed to be self-sustaining from the income generated in the provision of water services.

There are numerous factors that may affect the delivery of water services of a country's local authority. These include things like low water coverage, high level of unaccounted water, frequent water disruptions, inefficient revenue collection, high costs of water supply and distribution, poor water quality and other general mismanagement concerns. The ability of a water company to address these problems is one of the most important indicators to operational efficiencies. The following section is going to discuss some of these important indicators used in determining the efficiency of private companies in water service delivery.

Cost Recovery: The World Bank (2010) stated that, cost recovery is the ratio of revenue collected in relation to expenses incurred in operational and maintenance costs by a water company. Cost recovery is mainly used as a measure of business principles ability of ascertaining successful privatization of water services. In other words, sustainability of privatized water services depends on the ability of established water companies attaining a full cost recovery. It therefore means that an efficient revenue collection mechanism is a key step towards achieving full cost recovery. This would enable water companies to collect water bills from water consumers to cover costs incurred in production and supply of water services.

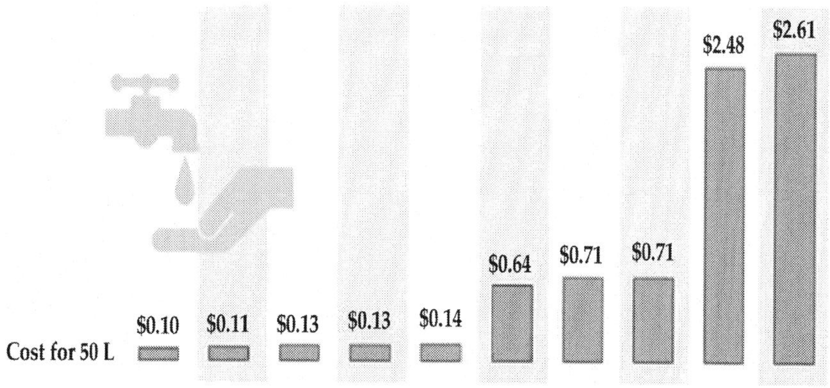

Source: https://www.statista.com/chart/4591/drinking-water-world-water-day/

Figure 7.5: The state of the world's water based on cost for 50 litres as of 2016

Operation and maintenance costs in the provision of water services include; personnel costs, energy costs, chemicals, network rehabilitation and expansion expenses. On the other hand, water company revenues are mainly sourced from the water bills collected from clients. The total amount a water company collects from the water bill is the maximum income a company can generate. However, companies still have other sources of funds such as grants, loans and donations. For a sustainable water service provision, a water company's total income should be in a position to meet its operation, maintenance and investment costs. Most countries (especially in developing regions) with established water companies are yet to achieve full cost recovery. This explains why most companies in these regions still rely on donors' and governments' subsidies, grants and loans for their operations and maintenance costs.

Privatization of water services delivery has seen increased efficiency in revenue collection in the established water companies. Increased revenue collection has been made possible by the capacity of private water companies to collect both outstanding arrears and current bills. The ability of water companies to maintain a positive trend in revenue collection is a means to achieve a sustainable water supply services. Increased revenue collection is usually attributed to privatization of water services in most countries. Most water companies are usually able to collect both outstanding and current arrears.

Privatization of water services also provides an opportunity for a review of water tariffs as water companies attempt to match their revenue with operation and maintenance costs. Water tariff review is based on a realization that water charges are the main sources of income for water companies to attain cost recovery. However, there have been concerns that an upward water tariffs would result in an increase in the cost of water. The resultant high cost of water would deny low income households access to water services. However, there is the need for setting regulations that directs water companies to make special provisions for low income households when reviewing water tariffs. Water companies should be a preferential water tariff review for water kiosks and vendors who are the main source of water for low income households in most places across the world.

There are several problems that may arise with the privatization of water service in a country. For instance, it increases cost of water which might be beyond reach for those who cannot afford. Increasing water tariff justified the attempts to meet full cost recovery which is the goal of privatization of water services. Water companies are supposed to institute cost cutting measures to achieve full cost recovery. These measures include reducing personnel costs and adjusting unit cost of production to match unit cost for a water tariff. Overstaffing and inefficient staff are major problems that usually face most public mode of water provision. Privatization of water services would inevitably lead to reduction of personnel costs through staff rationalization measures like retrenchment.

Almost half of operations and maintenance costs in private water companies usually go to staff remuneration-a clear indicator on high number of personnel. It is therefore the responsibility of water companies to develop a benchmark that water companies' personnel costs should not be more than 20%, 30% and 40% of operation and maintenances costs for large, medium and small water companies respectively. High number of personnel costs in water companies is attributed to unjustified hiring and high remuneration for the said staff.

Unit Cost of Production: Adoption of neo-liberal economic policies changed the conventional thinking that water is a social and an economic good. In other words, it is feasible to argue that there are costs incurred in the production and provision of water services, which are subsidized or offered by the government free of charge to the citizens. Access to water, as is access to food is a basic human right, and therefore just like the way access to food attaches some costs, the same should apply to water services. Considering water as an economic good that incurs expenses in its production and supply informs the introduction of a fee to consumers. This has been the major cause for opposing privatization of water services in most parts of world.

Privatization is the reason behind the introduction of water fee/charges, where there was none, removal of subsidies and increase of water tariffs in a bid to match production and supply costs is applied. For sustainable water supplies, water tariff per unit supplied to the consumers should match unit cost of production. However, available data indicates that the average water tariffs in different countries around the world have a propensity to be slightly lower than the unit cost of production. For instance, a country like Uganda in 2008 had national average tariff per unit of Ush 450 per cubic metre, while the unit cost of production was Ush 500 per cubic metre. This means that the water companies were under-pricing their service or making losses. This explains the water tariff review by the companies in a bid to match cost of production with water charges.

Unaccounted for Water/Non-Revenue Water: Non-Revenue Water has been one of the biggest challenges facing water sub-sector in most countries around the globe. It is against this background that level of Non-Revenue Water has come to be one of the ideal criteria in measuring operational efficiency of a water service provider. Unaccounted for water/non-revenue water is defined as the difference between the amount of water produced and amount of water supplied by a water company. Normally it would be expected that, when a water system is efficient and effective, the amount of water produced should be equal or relatively be the same as the amount distributed to consumers. However, the amount of water produced in many occasion is usually not the amount of water distributed to consumers. Some water is lost within the distribution network thereby compromising the capacity of water companies meeting consumer demands. High level of non-revenue water therefore plays a crucial part in denying access to water.

Source: http://jbswater.com/jbstravellogs.html

Figure 7.6: Illustrating a key Unaccounted for Water/Non-Revenue Water

It is worth to note that, the amount of water lost is almost equal to amount of water consumed. For instance, the cost of non-revenue water nationally in 2008 in Uganda (calculated at Ksh1200 per cubic meter), was estimated to be 6.8 billion Uganda shillings which was 25% more than the budget allocation to the water sector in the financial year 2008/09. This figure means that if unaccounted for water is reduced to bare minimum, water sector could be self-sustaining. Before privatization of water services in a country such as Ethiopia, unaccounted for water national average was at high of 80%, which has been reduced to 40% by the year 2009.

There are several causes for non-revenue water. These include illegal water connections, unbilled consumers, burst pipes, inaccurate master meter reader among others. Reduction of unaccounted for water has been attributed to efficient measures introduced to curb water losses. Most water company's revenue loss through non-revenue water are usually a major hindrance towards achieving the full cost recovery although it should be appreciated that privatization of water services can be used to considerably reduce high levels of non-revenue water witnessed in public water provisions. Since it is almost impossible to eliminate cases of unaccounted for water, the industry regulator should benchmark for water companies not to exceed 25%.

In a country like Uganda, privatization of water services has seen high level of unaccounted for water drop from 60% to 38% which is far beyond the recommended benchmark. This after the private water companies introduced a range of measures to remedy the loss of water. Some water companies even went further to acquire motorcycles for repair personnel to enable them address leakage problems promptly. Most water companies in Uganda also improved bill collection system by partnering with various financial institutions thereby addressing corrupt practices that were rampant in bill payments. However, there are water companies that still face challenges in reducing high levels of non-revenue water. Water companies should therefore institute measures and strategies to address high level of unaccounted for water. With a reduced level of non-revenue water, access to water is expected to increase.

Water Metering: Revamping dormant connections and ensuring all water connections are metered is one way of reducing high level of unaccounted for water. Metering connection enables water service providers to charge consumers according to what they actually consume. Unmetered consumers and the illegal water connections promote wasteful consumption of water thus, compromising the ability of water companies to meet their water demand.

The percentage of metering ratio, defined as the number of active metered connections compared to the total number of connections, is used to gauge the effectiveness of water companies in reducing the level of unaccounted for water. There are several advantages associated with metering of water supply systems. Some of these advantages include reducing the high level of unaccounted for water, increasing revenue collections, and widening the water company's area of coverage.

Water coverage by water companies is defined as, the percentage of population served, compared to total population living within the operation area of a water company. In most parts of the world (especially in developing regions) the national water coverage stands at 45%, half mark of the desired 90% although there has been an upward trend of about 5-10% in water coverage in some countries such as Kenya and Ethiopia which is considered very minimal. Investment in water sector through network expansion is considered a pre-requisite in increasing water coverage.

Source: http://www.wikihow.com/Read-a-Water-Meter

Figure 7.7: Water metering

Increased water coverage is attributed to the water company's strategy of including people served by water kiosks and yard tap in informal settlements where majority are low income households. Major investment in developing and extending water network in underserved and/or unserved areas and increasing metered connections can be a critical strategy of widening water company area of coverage. Increased coverage is attributed to water companies expanding water network and the establishment of more kiosks to serve low income households.

CONCLUSION AND RECOMMENDATIONS

Conclusion

Different scholars have argued that privatization can be used to introduce efficiency in provision of water services which translates to increased access to water services by consumers. However, there have been reservations whether low income households can still access water in a privatized water services regime. This thought is premised on the assumptions that, private water companies are unwilling to invest in low income residential areas which do not promise return. Failure by private water companies to invest in areas that do not promise profitable returns would therefore, compromise the availability of water services for low income households.

At the onset of privatization of water services, water companies are usually inclined to increase water tariff in a bid to attain cost recovery objective in their service provision. Increased water tariff consequently increases the cost of water which can compromise access to water for low income households.

Studies have shown that privatization of water services can help in introducing minimum level of operational efficiency in provision of water services. It has also been revealed that privatization of water services has positive impact on the availability of water services for low income households. A water market in the form of water kiosks and water vendors can be created to improve availability of water services for low income households. This can enable low income households access to water services at a recommended distance of one kilometre between a household and a water source.

The increased availability of water services for low income households is however negated by increased cost of water. Upon the inception of privatization of water services in a place, most water companies usually increase water tariffs in an attempt to attain cost recovery benchmark. In the same breadth, water kiosks operators and water vendors charge more than the recommended retail price as a justification to register profits. The increased cost of water for low income households is due to lack of an enforceable mechanism to regulate pricing of water services in low income residential areas. Unregulated water pricing can expose low income households to be exploited by water service providers through increased cost of water.

Recommendations

Privatization of water services in an area is an on-going process which has different impacts in different local authorities. The envisaged effectiveness and efficiency in water sector is far from being achieved in most parts of the world, especially in developing countries. A number of gains and challenges have been identified with regards to privatization of water service delivery. To strengthen the process and increase access to water, there are certain things that should be emphasized.

Lack of financial investment in the water sector can compromise the attainment of full cost recovery which is an indicator on the sustainability of private water companies. It is therefore important to have a continuous investment in the water sector to rehabilitate various water infrastructures, extend water networks so as to increase production capacity hence overcome the inefficiency of water supply disruption. These measures would greatly reduce level of unaccounted for water and increase water coverage in low income areas.

Retail price of water at water kiosks is beyond the recommended price that low income households cannot afford. It is recommended that enforceable mechanisms be adopted in order to ensure that water kiosks adhere to recommended retail price of water. The responsible water companies and regulatory bodies should work together with local communities in identifying water kiosks that flout price regulations. Issuance and renewal of operating license to water kiosks should be based on compliance with stated water pricing policy.

To further stabilize the price of water and augment the availability of water services for low income households, it is recommended that the water company reduces water connection fees and remove stringent conditions for low income household water connection. This would increase number of households with water connections and thereby stabilize water price that low income households can afford. Further research studies to evaluate privatization and its impact on access to water by categories of consumers comprising low, middle and high income households should be conducted.

8 MANAGEMENT OF NON-REVENUE WATER (UNACCOUNTED FOR WATER)

INTRODUCTION

Water provision services are the mandates of Water Service Providers (WSPs), or Water Utilities in many countries around the world. Their main mandate involves acquiring the water from the source, treating it to make it potable and fit for human usage, storing it, transporting it to the end-users, then billing the users based on usage tariffs and amounts used. This being a money-intensive endeavour, the WSP's depend on income from billing to carry out most of their functions.

The operations of these water utilities are however affected by inefficiencies, because not all of the water that they produce translates into revenue through billing. This unbilled water is termed as "non-revenue water" (NRW). Non-revenue water consists of both water provided for free to some consumers like firefighting departments, and water whose usage cannot be accounted for. Water utilised by firefighters and hydrants is easily tracked and logged through meters, and has been found to account for a very small percentage of non-revenue water. The bigger percentage of non-revenue water is thus caused by the unaccounted for water, or water losses.

Source: https://sensus.com/internet-of-things/smart-water/

Figure 8.1: Illustrating how non-revenue water is obtained

Water losses are classified into two main components; namely, physical and commercial losses. Physical losses are those caused by leakages of water throughout the utility's physical infrastructure, from the storage reservoirs to the pipes. Commercial losses are those caused by inaccurate billing caused by metering inaccuracies. Water losses are usually expressed as a percentage of the total water produced by the utility over a certain time period. These represents both a loss of revenue for the water utilities, and a lower level of service for their customers due to inadequate water provision.

There are several ways through which non-revenue water losses can occur. These include lost revenue for the utility companies; the reduced level of service as there is less water to supply to consumers; and reputation loss for the utility companies due to failure to satisfy water demand.

The customers of the water utility also suffer because of these water losses. Since there is less water being supplied, many of them are usually unable to get adequate water for their needs from the water utility companies. This leads to them seeking alternative water sources such as donkey vendors and water bowser suppliers, which are expensive and whose water is of dubious quality standards. In addition, the water they get from the utility is more expensive as the utility tries to recover its losses by overcharging the paying customers in their bills. Non-Revenue Water therefore has a steep financial impact on water utilities and their customers, and needs to be studied extensively and managed effectively in order to reduce its negative effects.

When conducting any research study that pertain non-revenue water losses, there are several points that must be considered. First, it is important to find out the extent of non-revenue water that is experienced in an area. Second, it is important to establish the likely causes of non-revenue water and classify them based on the nature they occur (physical or commercial losses). Thirdly, it is important to assess the effects that are likely to occur due to the losses on both the utility water companies and to the end customers.

WATER PROVISION PROCESSES

According to World Health Organization (2010) report, it has been concluded that only about 40% of people worldwide have direct access to tap water. The rest, usually the poor in informal settlements, usually get their water from illegal piped connections, water kiosks, and water vendors.

Some people have access to private means of water supply, and others also sink boreholes to supplement their water supply. However, water supply is erratic, and only about 40% of them receive water on a 24-hour basis. To achieve their mandate, the water utility companies have the responsibility to source the raw water, transport it to a treatment plant, treat it to make it potable, and store it in reservoirs and tanks, from where it is distributed to consumers via pipes. This is then followed by the billing process, involving meter reading and revenue collection. On the other end, the utility also collects wastewater through the sewer line, and treats it before releasing it to the environment. These steps are outlined in detail below.

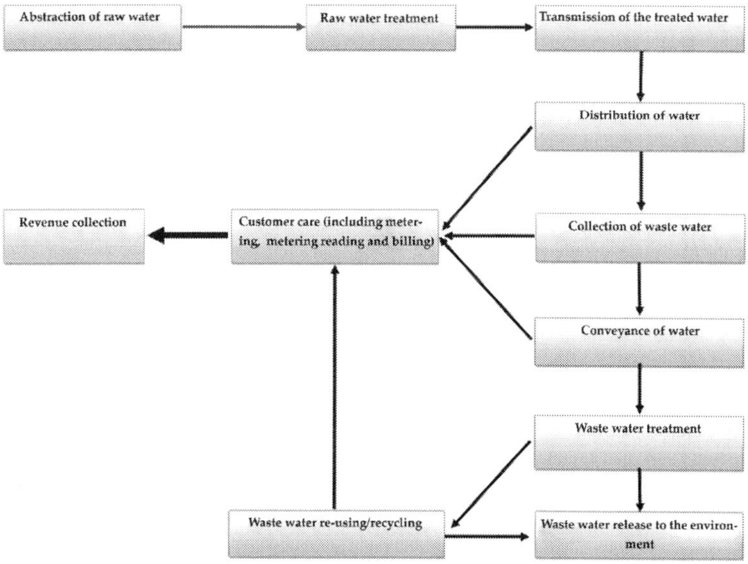

Source: http://slideplayer.com/slide/6969071/

Figure 8.2: An overview of water utility company's main functions

Sources of Raw Water. There are two key sources of water that is supplied by utility companies to the end customers. These sources are the surface and underground sources. Some utility companies in different parts of the world rely wholly on surface water, while the underground water is accessed through boreholes by mostly private entities, some of whom have access to the main water supply from the utility companies but also use the boreholes, especially as a backup source of water during to erratic supply. Surface water is mainly sourced from the river basins, from various reservoirs, dams and other surface water sources.

Treatment of Water: Water obtained from the various sources is then taken to a treatment plant. Every country usually has a mandate to meet the minimum World Health Organization water standards for what constitutes a "potable water". In other words, the World Health Organization (WHO) guidelines on water quality form the basis for water standards supplied by the water utility companies in different countries around the world. There are several parameters that must be evaluated with regards to the quality of water. These parameters include;

- Turbidity: turbidity measures the extent to which the intensity of light passing through water is reduced by the suspended solids in the water. Water should have an acceptably low level of turbidity.
- Alkalinity and pH: this is a measure of the pH of the water. Water should have a pH close to 7 (neither acidic nor alkaline).
- Suspended solids: solids are measured as residue from evaporating the water. They consist of dissolved and suspended solids, and must be kept to a minimum after treatment.
- Dissolved oxygen (D.O.): this is a measure of the concentration of oxygen in the water.
- Hardness – this is the property of water that prevents lather formation with soap. There is carbonate hardness and non-carbonate hardness.
- Residual chlorine, chlorides, irons and other ions – ions are dissolved in water from the environment.
- Bacteria: this involves the determination of possible fecal contamination of water.

To achieve the above standards, a water treatment plant generally subjects the raw water to many treatment processes. The processes generally consist of:

Physical processes: these are used to remove solid matter in the water. These include sedimentation, screening, filtration, and coagulation and flocculation.

Chemical processes: these are employed to remove the chemical matter not removable by physical means, like manganese, iron, and hardness. They include chlorination, addition of lime, and alum. In doing all the above treatments, the water utility incurs costs. Such costs are from power charges, purchase of the chemicals, and maintenance of the equipment/machinery in the treatment plant.

Storage and Transportation of Water: The other major element of the water provision process is the storage and transportation. Treated water is stored in reservoirs and tanks which are constructed within the treatment plants. The major challenge facing storage facilities is siltation, which decreases their capacity. This capacity reduction results in increased incidences of water overflows, leading to loss of water. Regular desilting is done at the cost of the water utility.

In all the processes, water should be transported, whether raw water from the source to the treatment plant, or processed water from the treatment plant to the final storage tanks and reservoirs. Fortunately, in some countries, water sources are usually situated at a higher altitude than the end users. This means that water flows in the pipes freely by gravity without requiring the use of pumps, which saves a lot on costs. The main challenges facing transportation include aging pipes, which are now prone to leakages and bursts due to weakening over time. The pipes also suffer from sedimentation and other sources of impurities, which affect the quality of the water. Some water utility companies in different places around the world usually suffer from improper mechanism of promptly locating and fixing bursts, which leads to excessive losses of water during bursts. Poor pipe security means that along the pipeline, people easily tamper with it to form illegal water networks, especially near informal settlements. Such networks affect the quality of water in the main pipe, while also reducing the pressure of the water required for delivery to customers further on along the grid.

The other major function in transportation of water is distribution. Distribution networks are designed by engineers to ensure that all regions are served based on water amount requirements (calculated based on population and the income level in the area), adequate pressure to serve areas with many high-rise buildings, and the relief characteristics of different areas, for instance in some areas which are higher than the source location of the water, enough pressure (head) must be provided to ensure the water reaches the people. Distribution requirements also affect parameters such as pipe diameters and material.

Billing and Revenue Collection. Billing of the water to consumers is the main way water utilities recover their costs. In almost all water utilities in the world, water is a post-paid service. Customers are billed according to how much water they have used. Water usage is determined by meters, which are normally placed at the sites of the water usage (mostly residences). A utility worker goes round reading the bills and recording them monthly. These are then compared to the previous readings and the difference is the usage for the month. In practice, sending people to read bills monthly is difficult, so the utility estimates usage based on previous patterns. The discrepancies between the estimates and the actual usages are then reconciled periodically.

Once the usage has been determined, the bills are typically sent via post. The customer then pays a tariff, which is based on the usage. The bill amount typically increases as more the water is used. This progressive billing ensures that heavier users pay more as they tend to be better-off financially, reducing the pressure on the poorer consumers. If the residence has a sewer connection, this is billed based on the usage of the water, typically at a rate of 75% of the water usage.

Sewerage Services: In addition to water provision services, water utility companies also provide sewer services. Not all buildings with water connection are connected to the main trunk sewer line, which is due to regulation gaps in a county building approvals process, and reduced capacity of the utility companies to provide sewers in all locations. From the buildings, raw water is transported to either of the main wastewater treatment plants. The treatment plants focus is on reducing the biochemical oxygen demand of the water to sufficient levels as agreed with the responsible Environment Management Authority within a country before disposing the waste into the environment.

Wastewater treatment has many similarities with treatment of water to make it potable, the difference being that the initial quality of wastewater is so poor that it would be too costly to treat it until it is good enough for human use. Therefore, it is discharged into the rivers, as a way of recharging it into the natural hydrological cycle.

UNACCOUNTED FOR WATER

Most utility water companies around the world suffer from financial challenges, despite the fact that all the water they produce seem to be used up. It has been established that most of the water produced usually do not end up bringing in revenue for the utility water companies, even though they are billing their customers monthly. The International Water Association, IWA, termed this water that could not be accounted for as 'water losses'.

Non-revenue water is all the water that has been produced but has not earned any revenue for the utility. Utility water companies are required to provide water free of charge for some services of social good like firefighting services, and to have fire hydrants with adequate pressure always ready. Because this water does not earn any revenue for the utility, it is also non-revenue water. However, studies have shown that most of the non-revenue water is caused by water losses, and that the water provided free of charge is normally insignificant in its impact on the revenue collection levels.

According to IWA, there are different sources through which water can be lost. These losses can be tracked using a number of balance techniques. In other words, these strategies are essential to water utility companies in tracking their entire distribution systems so as to discover where the losses are occurring. The following table illustrates standard components that can be used by water utility companies in comparing their water provision services with those of the World Health Organization. These standard components can be used to determine the level of discrepancies that may arise with regard to water provision processes.

Table 8.1: The International Water Association Water Balance Table

<table>
<tr><td rowspan="10">Water Input</td><td rowspan="4">Authorized consumption</td><td rowspan="2">Billed consumption</td><td>Billed metered consumption</td><td rowspan="2">Revenue Water</td></tr>
<tr><td>Billed unmetered consumption (flat rate)</td></tr>
<tr><td rowspan="2">Unbilled consumption</td><td>Unbilled metered consumption (billing error)</td><td rowspan="8">Non Revenue Water</td></tr>
<tr><td>Unbilled unmetered consumption (eg firefighting)</td></tr>
<tr><td rowspan="6">Water Loss</td><td rowspan="3">Apparent losses</td><td>Customer meter inaccuracies</td></tr>
<tr><td>Data handling errors</td></tr>
<tr><td>Unauthorized consumption (theft)</td></tr>
<tr><td rowspan="3">Real losses</td><td>Leakage on mains pipes</td></tr>
<tr><td>Leakage and overflow on tanks</td></tr>
<tr><td>Leakage on service connections</td></tr>
</table>

Source: http://www.scielo.org.za/scielo.php?script=sci_arttext&pid =S1816-79502009000200005

****Kindly retype this table using Microsoft, this is a figure in its present state. The use the open table format.**

Water input system: this is the annual (or other period) total water produced and put into the utility's distribution networks. It is composed of

- Authorized consumption: this is the amount of water whose end usage was what is intended (authorized) by the water utility company. The water in this component may be billed or unbilled, and metered or unmetered. It consists of:
- Billed authorized consumption: this is the water which was billed, and brought in revenue for the utility water company. It can either be metered or unmetered. Metered consumption is where the meter was read, and unmetered consumption is usually due to the usage of estimates in calculating the water bill. Revenue water is equal to this billed consumption.
- Unbilled authorized consumption: this is all water consumption, metered or unmetered, for which the utility has authorized usage without charges. This is usually in the case of provision of water for fire hydrants and other public services. It is part of non-revenue water, but normally accounts to an insignificant percentage of it.
- Water losses: this is the component of the total input volume whose usage was not according to the intentions of the water utility. It is also called the unaccounted for water, or unauthorized usage. This portion contributes the most to non-revenue water, and is divided into:
- Commercial losses: these are all losses that occur due to metering issues. They are also termed as 'apparent losses', because the water is found to have gotten to the end user, but there was no payment for it. These losses are usually caused by metering inaccuracies (where the meter underestimates the usage), meter tampering, and outright water theft along the distribution line. This mainly happens near slum areas (informal settlement areas), where the utility's pipes are exposed and it is difficult to properly secure them.
- Physical losses: these are all the losses of water from the distribution network. They are considered to be 'real' losses, because they are caused by physical problems in

the infrastructure. They consist of losses in the storage reservoirs, either due to overflows or leakages, and losses in the distribution main pipes caused by leakages and bursts. A certain component of these losses also comes from customer distribution networks before the metering point, due to leaks and sometimes poorly done pipe joint connections.

Considering the standard practices, non-revenue water is expressed as a percentage of the total water produced. From the above table, a simple equation to calculate the level of non-revenue water in any water supply the system can be expressed as follows;

$$Non\,Revenue\,Water = system\,input\,volume - billed\,authorized\,consumption$$

$$= total\,water\,supplied - water\,solid$$

Expressed as a percentage,

$$non\,revenue\,water = \frac{total\,water\,supplied - water\,solid}{total\,water\,supplied} \times 100$$

WATER BALANCE COMPONENTS

Water balance is generally accepted to define water losses. However, according to studies, water balance is more applicable in the developed world since most of the revenue billing for water services is collected. But in the developing countries, a substantial amount of billing is never collected and therefore may be considered as losses.

Calculating the water balance component or what is commonly referred to as water auditing is of the key strategies that are used in reducing water losses with a supply system. To track the losses, data is collected from the water supply network and the various components of the water table are determined. Tracking of water usage requires the utility to perform extensive data collection throughout its network, as well as using estimation methods to determine the other components of the water balance table to a reasonable level of accuracy. The indicator of non-revenue water as a percentage of total system input could be misleading, because not all losses are the same. There are certain steps that must be followed when calculating or determining the losses of water. The steps include:

Determination of the System Input Volume: This involves identification of all the sources of the water that the utility water companies supply. In most places around the world, this is the total output of water from the treatment plants. Meter accuracy must be ensured by checking the manufacturer's manuals, and setting up additional meters downstream in order to confirm the readings. The data collected should be subjected to statistical analysis and confidence limits of 95%.

Determination of Authorized Consumption: The billed metered consumption can be determined from the consumption indicated by the meter readings. Allow for meter reading time lag (because the reading is always done after consumption). Unbilled metered consumption should be determined in the same way as billed metered consumption.

For the unmetered consumption, the extracted usage data from the connections serve as a good estimate. Some of these connections could also be installed with meters in order to get more accurate figures.

Unbilled unmetered consumption should also be carefully estimated. This component usually involves water used by the utility company for flushing its water systems e.g. desilting of tanks. The water used should be carefully logged. Many water utility companies tend to overestimate this portion so as to make water losses seem less than they really are. This is because it is part of non-revenue water.

Estimation of Commercial Losses: Estimating unauthorized consumption is a difficult task. The best way of improving the accuracy of such estimation is intimate knowledge of the distribution network. This can be done through regular surveys, sampling of portions of the network, and using statistical methods in order to improve the confidence limits of the estimations.

Meter inaccuracies and data handling errors can be estimated through sampling of the meters. Use of software is useful in this regard because software can detect when usage patterns vary widely from the expected. Faulty meters should also be promptly replaced. Data handling errors can also be handled by using database software, so as to detect discrepancies. Such errors may arise from mistakes during input by the meter reader.

Calculation of Physical Losses: This is determined using the formula:

$$Physical\ losses = result\,1 - result\,2 - result\,3$$

Estimating Physical Losses Components: to accurately determine each component of physical losses, a detailed component analysis of the system should be done. This can be preceded by estimates.

- For leakages on distribution mains, the reported bursts can be used based on the repair records. The invisible leakages, for instance, that occur slowly over a long time, can then be estimated based on parameters like the ages of the pipes. The amount of water lost in a leak is the product of the leak flow rate and the leak duration.
- For leakages and overflows from the storage tanks, this component is easily quantifiable if the utility logs all its data. This is done using level drop tests when the outflow valves are closed.
- For leakages on connections up to the customer meter, these can be estimated by taking the physical losses determined above and subtracting the mains and storage facility leakages.

KEY EFFECTS OF NON-REVENUE WATER LOSSES

Water losses affect all the water sector stakeholders, namely the water supply companies and their customers.

Effects on Water Utilities: All businesses survive on sale of their marketable products (water). If a big portion of it is lost, the business loses money and its capacity to produce its marketable product reduces. This leads to lower quality of the product and less of it produced, which further affects profitability. In water utilities, the marketable product is treated water. The above challenge leads to a vicious circle, as illustrated below:

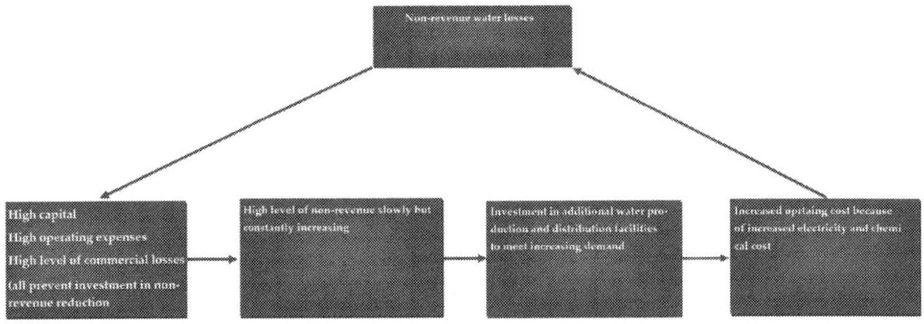

Source: https://thewaternetwork.com/_/integrated-water-resource-management-iwrm/article-FfV/water-losses-non-revenue-and-unaccounted-for-water-raoNh8dY0nlBOYV0qfLbpA

****kindly consider enlarging the figure below to make the writings more readable. They are too small to read**

Figure 8.3: The vicious non-revenue water cycle

If left to occur at a high rate, non-revenue water leads to reduced capacity of the water utility to perform its functions due to the worse financial position. This means that functions like water abstraction, treatment and transportation are impacted. Water utilities require money to run and maintain their infrastructure like pipes and machinery at the treatment plants. With the diminished finances, the water utility company is less able to do this, leading to a reduction in the amount of water it can treat, or poor quality of its treated water, or both. This means that its capacity to serve its customers is reduced even further. This leads to even more non-revenue water losses, and the vicious cycle goes on and on.

Effects on Customers: The main objective of a water utility company is to satisfy customer demand. In most places around the world, especially in developing countries, the total system input volume is not enough to satisfy all the water demand for the citizens. If non-revenue water is added to this, it results in an even more diminished capacity to provide water to its customers. The main effect of this diminished capacity to satisfy demand is an intermittent water supply. For instance, in a country like Kenya, it has been indicated that only 40% of the customers receive water on a 24 hour basis.

Source: http://pulitzercenter.org/reporting/pathway-peril

Figure 8.4: The problem of water supply impacts the poor the most

Erratic water supply disproportionately affects the poor. Most of the zones with 24 hour supply are the high-income zones. Areas which are known to attract the lower middle class are highly affected by the erratic supply. However, in poor informal settlements there is almost no piped water supply to households. Most of the people depend upon donkey cart vendors, who sell water in 20 litre jerry cans. This water is sold at disproportionately high prices. For instance, in Kenya, a 20 litre can of water going for Ksh 10 translates into Ksh 500 per cubic meter of water, this is way higher than the standard utility company price of about Ksh 12 per cubic meter. This is being solved by the emergence of water kiosk vendors, but there is still a long way to go.

Erratic water availability affects customer relations with water utilities, leading to less willingness to pay. In addition, the water utility companies, on trying to recover losses from non-revenue water, usually resort to higher tariffs for their water supply. This affects paying customers unfairly, as they subsidize the rates for those stealing the water. These conditions further contribute to the vicious non-revenue water circle. The urban poor are normally blamed for theft of water from main pipes through the establishment of illegal networks. However, because they do not have water intensive appliances like washing machines, the poor tend to have a much smaller effect on non-revenue water than previously thought. The highest incidences of commercial losses, studies show, are caused by meter tampering and theft by the more well-off residents, and industrial users who typically use a lot of water.

NON-REVENUE WATER REDUCTION AND MANAGEMENT STRATEGIES

Non-revenue water reduction strategies should always be carefully formulated to ensure their success. Most reduction strategies are money intensive. However, their main justification comes from the fact that they tend to pay off in a big way due to the virtuous non-revenue water cycle.

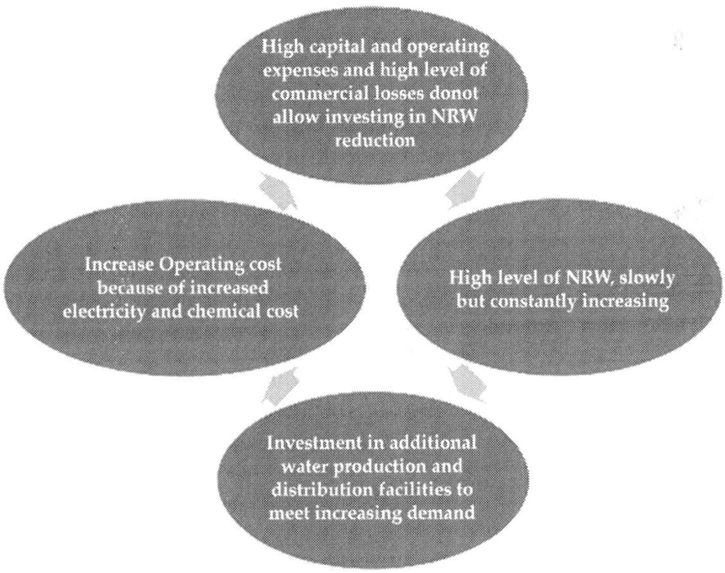

Source: https://www.slideshare.net/IPPAI/dr-chandra-mohan-b-iasmanaging-directorchennai-metropolitan-water-supply-and-sewerage-board

Figure 8.5: The virtuous non-revenue water cycle

According to several studies, efforts to reduce non-revenue water lead to an increase in financial capacity for the water utility, which leads to better services and more income, which can further be invested in more non-revenue water reduction, leading to even more benefits. The benefits are shown in Figure 8.5 on the virtuous circle form the main justification for a water utility to invest its money in a non-revenue water reduction strategy.

Determining Non-revenue Reduction Strategy

A proper non-revenue water management and reduction strategy can only be formulated after the causes of non-revenue water are determined, and their components formulated. This can be done by performing a water audit, as demonstrated previously. Establishment of this strategy entails:

Establishment of the Strategy Development Team: A non-revenue water strategy team should first be constituted. The approach should be holistic, and should consist of members from all departments in the utility company, from the customer service, operations department, engineers, and finance departments. Many utilities usually have a dedicated non-revenue water department. While this is prudent, the other members of the organization tend to view non-revenue water as the department's problem, rather than the whole utility.

***Setting Appropriate Reduction Targets*:** According to studies, the best way to manage a problem is not necessarily to eliminating it, but by reducing it to acceptable levels. Solving problems requires funding, and while a theoretical value of 0% non-revenue water may look optimal, it may be too costly to obtain. Establishing appropriate levels of losses should be done using the available resources considering the constraints and other functions of the utility. The following is a graph that can be used to obtain what is known as the economic level of non-revenue water.

As shown in the graph, water utility companies should use the cost of water lost (which increases as the level of non-revenue water increases) and the cost of non-revenue water reduction (which reduces as non-revenue water increases). Non-revenue water reduction costs proportionately more as the non-revenue water level reduces.

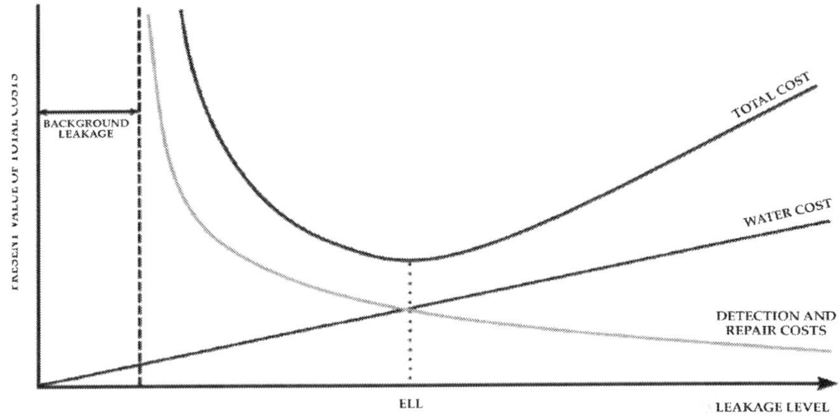

Source:https://www.researchgate.net/publication/46419700_Incorporating_energy_use_into_the_economic_level_of_Leakage_Model

Figure 8.6: Illustrating an economic level of non-revenue water

Theoretically, it requires infinite resources to reduce non-revenue water to 0%, which is not feasible. The economic level of non-revenue water is the point where the two curves intersects, representing the maximum non-revenue water reduction for the minimum cost incurred.

***Prioritizing the Non-revenue Water Components to Reduce**:* In non-revenue water reduction, the best strategy is to begin with the easiest components to reduce, or the ones that bring the most immediate returns. Using the water table audit, the team should be able to identify the best components to reduce first, taking note that: Physical losses – solving physical losses results mainly in the saving of revenue and reduced operation costs. However, solving physical losses generally requires more revenue investment, both in detection of where the losses are occurring and the costs of repairs. However, due to the relatively high volume of water lost in this way, reducing them has the greatest impact on the percentage levels of non-revenue water.

Commercial losses – these losses, while representing a lower volume of water, tend to represent the highest levels of revenue achieved. Solving a problem like faulty meters generally results in immediate returns, because the consumer begins paying fully as soon as the next billing cycle.

It is up to the team to do all the analysis and to come up with a program on the components to focus on, considering the results expected in solving each.

Awareness, Location, and Repair in Non-revenue Water Reduction: once the individual reduction activities have been determined in the step above, the strategic team should use the concept of awareness, location, and repair as it carries out the activities identified. All non-revenue water losses have these three components.

Awareness – this is the time required for the utility to become aware of the loss, whether in the form of a leak, or a faulty customer meter.

Location – this is the time required to locate the problem, e.g. finding out where the leak is occurring, or where the faulty or tampered meters are.

Repair – this is the time required to mobilize the resources and repair the problem.

The three components should be kept at a minimum; because the more time the problem remains unsolved, the higher the volume of the losses incurred:

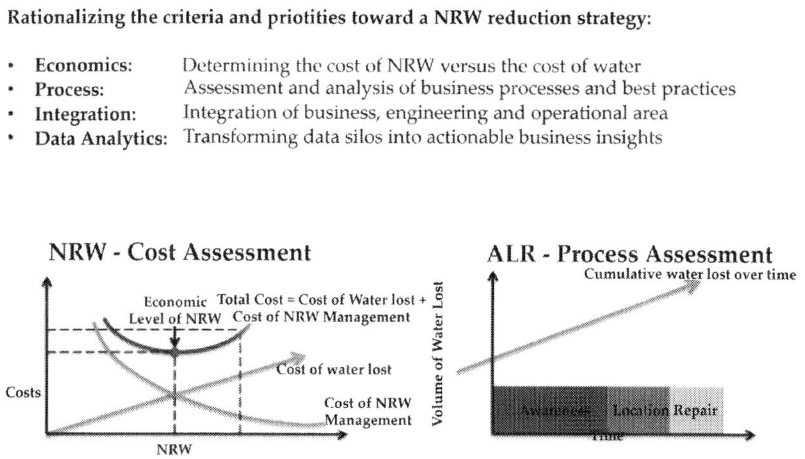

Source: https://www.slideshare.net/globalwaterpartnership/running-dry-smart-water-and-leak-detection-by-steven-windsor-17799034

Figure 8.7: A more detailed illustration of the economic level of non-revenue water

Budget Considerations for the Strategy Implementation: The overall cost of reducing non-revenue water could be very high as it may take many years. Strategies tend to start off at full momentum only to suffer budget cuts over time. It is always best to discuss a long-term budget with the stakeholders before starting. Conducting pilot projects to demonstrate the effectiveness of various strategies should also be done to convince the management.

Management and Reduction of Commercial Losses

Commercial losses are typically losses of revenue rather than the water itself. This means that the water has been used, but water utility company does not receive adequate compensation for it. This happens due to water theft and faulty meters. Thus, commercial losses are also known as apparent losses.

Reducing of commercial losses has a greater value because revenue increases are realized almost immediately. Reduction of commercial water losses can be classified into four main areas, namely; customer meter inaccuracy, unauthorized consumption, meter reading errors, and data handling and accounting errors.

Addressing Customer Meter Inaccuracy: Inaccurate meters tend to under-register the consumption. Over-registration of water consumption rarely occurs due to faulty meters and in such cases, customer reporting of the issue is prompt. In solving this problem, water utility companies should focus first on commercial and industrial users, who use higher amounts of water and pay higher tariffs. This is where the most returns can be achieved. The main strategies employed to solve such inaccuracies are:

- Proper installation and handling of meters – meters should always be installed properly, as per the manufacturer specifications. Some meters require a certain length of pipe flow upstream or downstream for higher accuracy. Utilities should take it upon themselves to source for trustworthy meters and install them for the customers, in a place where they can be easily read. In a multiple residence area, meters should be properly numbered according to the property they serve, to avoid errors during meter reading.

In addition to this, meters should be sized properly, and the proper class and type must be used for every situation. For instance, some larger meters do not register flows of water if the flow rate is below a certain threshold, and should only be installed in parts of the network where the flows justify them. Meters are also classified based on how sediments in the water affect their accuracy. If the water quality is poor, meters whose accuracy is not affected significantly by sediments should then be used.

- Monitoring water quality and intermittent water supply – water quality has a great effect on water meters, especially those sensitive to sedimentation. Utilities should regularly monitor their water quality to reduce this. Intermittent water supply, while sometimes necessary in the case of water rationing, tends to reduce the accuracy of meters due to air gaps in the pipes.
- Maintenance and repair of meters – meters should be routinely maintained and replaced. Systematic replacement should be done regularly, beginning with the older meters first. Meter maintenance includes examining the mechanical parts, washing off sedimentation, and calibration to ensure accuracy.
- Addressing meter tampering – some customers insert objects into meters to under-register their usage. Therefore, water utility companies should consider using the more tamper-proof type of meters in the market.

Addressing Unauthorized Consumption: Unauthorized consumption involves mostly illegal connections, meter bypassing, and errors in the billing system.

Illegal connections – this is the physical installation of a water connection in the distribution pipeline without the knowledge of the utility. They are normally installed at pipe joints. Such connections also lead to the problem of pressure tampering in the distribution system, leading to further inefficiencies. Customer awareness on the effects of such connections is important, and they should be encouraged to report such incidences. In addition, the water utility should regularly take rounds auditing their pipe networks to detect such connections and rectify them.

Meter bypassing – some customers install meter bypasses, meaning that they use water without getting metered (this is different from tampering to reduce meter accuracy). Bypass pipes are often hidden underground, and is done by corporate users who then use only a little of the metered water while the rest goes through the bypass pipe. Water utilities should always conduct flow balance analyses, to see whether the bills being paid correspond to the usage levels. This leads to early discovery of the culprits.

Billing system errors – sometimes consumers are connected to the grid but billing departments are not notified to add them to the billing list. Utilities should regularly conduct customer surveys and reconcile the lists to capture these consumers into their systems.

Corruption by meter readers - this also contributes to the errors. Some meter readers conspire with customers to record a lower usage. One way of reducing this is preventing familiarization of the meter readers with the customers, by ensuring their routes are changed periodically.

Addressing Meter Reading Errors*:* These errors can be caused by negligence, aging meters, and corruption. Sometimes the meter readers make errors while taking their readings, or guess values due to low visibility of the meter dials. Meter readers should always be properly trained and motivated so that they can form a better link between the utility and its customers. Frequent spot checks should also be done, and corruption should be punished severely as deterrence. Conspiracy between meter readers customers should also be dealt with through fines or service suspension.

Addressing Data Handling and Accounting Errors*:* The bill reading process consists of the meter reader taking the readings, recording them, and then taking the readings to the billing office, which inputs the values into the system. Distribution of consumption reports and billing must be done efficiently. As such correct billing statement must be delivered to the correct recipients. A robust billing database should be used to easily detect errors, especially when it notes a big variation in the amounts and the expected usage for the customers. Updating the database also with recent zoning changes or modifications on establishment or housing numbers must be done to avoid erroneous billing deliveries.

Utilities also regularly send estimates to consumers, especially where meter reading is not able to be done periodically. Over time, this system develops errors because the estimates are only based on confidence levels. Therefore, water utility companies should periodically regularize their readings by sending actual meter readers, to reconcile the readings to the actual values.

Management and Reduction of Physical Losses

Physical water losses are also termed real losses. This is because they involve real loss of water from the utility's distribution system, and not misappropriation of the water usage as in the case of commercial losses. Physical losses reduce the amount of water in the distribution system.

Reduction of physical losses thus results in an increase of the volume of water in the system, and do not immediately result in revenue. However, the amount of water lost is usually very significant, and should be considered because the water utility needs more water to satisfy the demand. Physical losses are classified into three main components:

- Leakage from transmission and distribution mains – these are usually large events, and get noticed faster. They usually lead to disruptions in the infrastructure, but due to their relatively quick detection times, they are shut off and repaired faster.

Source: http://keywordsuggest.org/content/640838-water-burst.html

Figure 8.8: A mains water pipe burst

- Leakage and overflows from the utility reservoirs and storage tanks – these leaks are easily quantified. Overflows should be observed by the utility managers so that their levels can be estimated. Leakages from tanks can be done by drop level tests by measuring the rate of drop in water levels after shutting off inflow and outflow from the tank.
- Leakage on service connections up to the customer meter – unlike main pipe bursts, these are generally harder to detect as they are not easily visible. They tend to result in the highest losses because they go on for longer time period. They can be estimated by deducting the leakages from the mains and the storage facilities from the total physical losses.

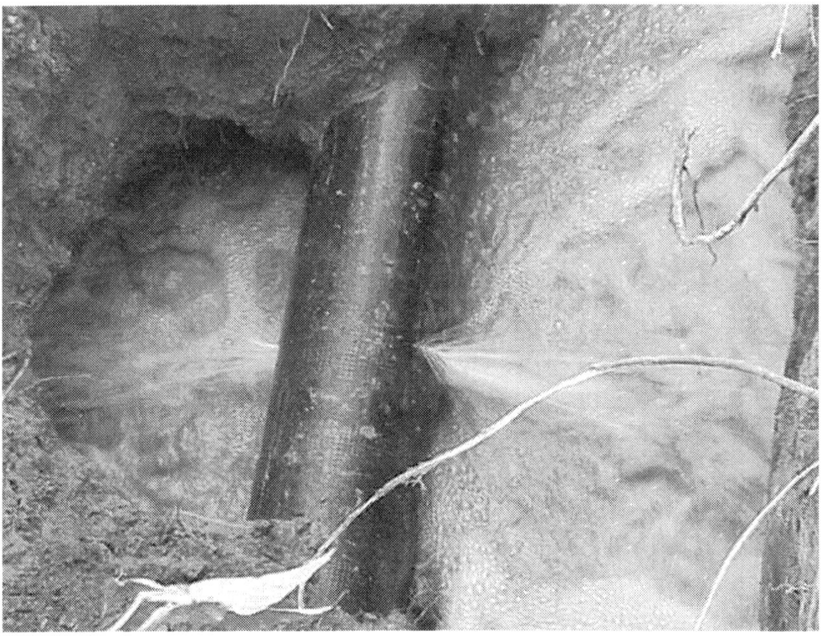

Source: http://www.allianceforwaterefficiency.org/Water_Loss_Control_-_What_Can_Be_Done.aspx

Figure 8.9: Leakage on water supply service connections

All physical water losses are thus leakages; the only thing that varies is the location of the leakage in the system, and their severity and duration. Physical loss management is thus leakage management, but the various characteristics of leakages should be understood to come up with a proper strategy to combat the losses.

Based on the ALR (Awareness, Location, and Repair) technique, service connection leakages, through having lower flows, have high losses due to slow detection levels. Mains leakages have higher flows, but due to the faster detection, are stopped before they lose much water. There are three basic types of leakages:

- Reported bursts – these are visible, and promptly reported by the public. They have a short awareness time.
- Unreported bursts – these mainly occur underground, and are only discovered during leak detection surveys. They have a long awareness time.
- Background leakages – these are small leaks, often numerous, that are not cost-effective to repair individually.

Non-revenue water managers should thus be aware of these leakage types before coming up with a strategy. There should be active programs to detect and deal with the leaks. There are several components that can be used in the management of leakages along water supply system. These components include;

Active Leakage Control (ALC): This involves actively monitoring the system, even when leaks are not reported. Analysis of the distribution network involves a technique called *flow metering*. This technique captures bursts quicker, and enables estimation of small background leakages. It involves data logging and analysis via Supervisory Control and Data Acquisition (SCADA) systems, which utilities use in their networks, especially the District Meter Areas (DMAs). After the leaks have been noted, the next step is locating the exact location for repairs. This is done using various techniques, including:

- Noise loggers – these narrows down the localization of the leakages by analyzing the noise characteristics of the pipes. They are placed at intervals in clusters, usually 6, 12, or 18 loggers. Noises that are suspected of being caused by leaks are then noted.
- Leak noise correlators – these instruments locate the leaks using the velocity of sounds rather than the noise levels.

- Ground microphones – these typically amplify the sound of a leak to make it easier to locate.
- Sounding sticks – these are like doctor's stethoscopes. They consist of a simple rod with a hollow ear piece attached. The logger then listens for any leakage noises. The staff members must be trained to be accurate in doing this.

Source: https://www.pinterest.com/h2onet_co_za/water-leak-detection/

Figure 8.10: Use of a sounding stick to detect water leakages

***Pressure Management**:* Pressure affects leakages in two ways: it increases the likelihood of a burst, and if a burst occurs, higher pressures lead to higher leakages. The relationship between pressure and leakage is thus, one of direct proportionality, i.e. a linear relationship:

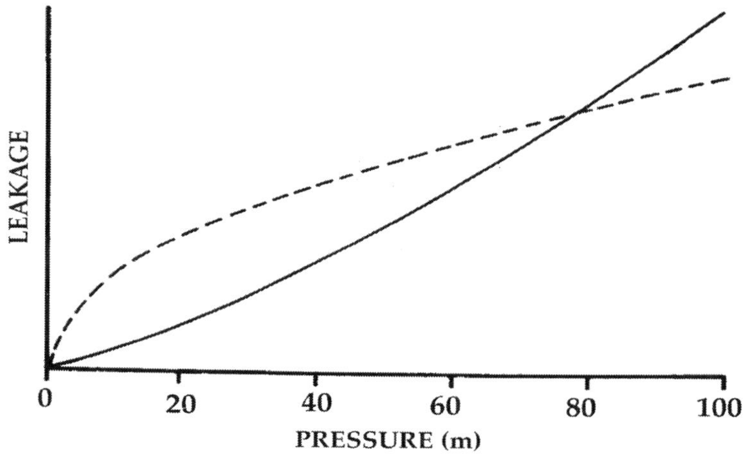

Source: http://www.sswm.info/content/leakage-control

Figure 8.11: Pressure and leakage relationship

Water utility companies must first analyze the needs of the system before embarking on pressure control. Some of the issues to look at includes the customer types and needs. High-rise buildings require higher pressure due to water supply in higher floors. After gathering this information, the utility company should then model the pressure control system, whether using software or in a pilot project section, to clearly understand the costs and benefits. Once this is done, the pressure control apparatus is then installed at the correct points. One of the most cost-effective apparatus for pressure control is the automatic pressure reducing valve (PRV). These vales maintain the same downstream pressure, regardless of the upstream pressure levels.

***Speed and Quality of Repairs**:* Once leaks are detected, water utility companies should ensure they are repaired promptly. Delays, as the ALR analysis shows, lead to accumulation of losses. The key to prompt repair is efficient organization, sufficient funding, diligent management and staff, and good quality of service connections.

***Asset Management**:* Proper asset management is the key to reducing losses. The assets of a water utility include the entire infrastructure in the whole network, from the reservoirs, pipes, and water meters. A proper asset management policy does the following:

- Understand their assets – the utility should always know their assets, in terms of their condition, number, materials, and how they are performing currently.
- Data collection – data should be collected on all the assets to determine how well they are performing. It should also be done on the environment and the market. If a better way of doing things comes about, whether due to innovation or changing standards, the utility should be able to come to a decision.
- Good information systems – the best way is through computerized logging of their assets. This means that information on when to do replacements, which components in the market to replace them with, and other considerations can be done to make the non-revenue water (NRW) management easier.

Non-revenue Water Management using District Meter Areas (DMAs)

Water utilities that operate their pipe networks as an open system, where water from many sources is fed into the system and mixed, suffer from problems of water pressure and quality. This leads to many utilities dividing up their networks into smaller zones, called district meter areas (DMAs). These DMAs make it easier to manage the system in terms of pressure and water quality levels. Each DMA consists of its own intake from the main system, and meters and other data loggers for performance analysis.

It has also been noted that the benefits of DMAs go beyond the network improvements, but also make non-revenue water management easier. It is easier to analyze a smaller system and to determine the losses in each, and their causes and mitigation. This means that non-revenue management could be undertaken actively. Zones may follow either administrative or hydraulic boundaries.

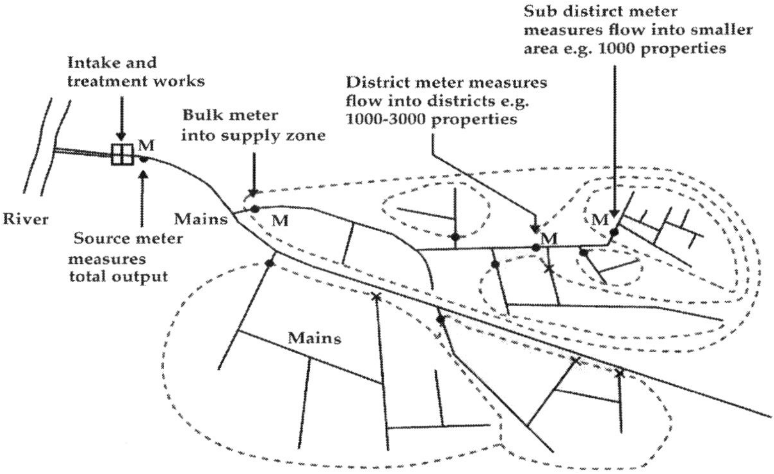

Source: http://www.sswm.info/content/leakage-control

Figure 8.12: Typical DMA layout

District Meter Areas make it easier to isolate problematic areas, determine and reduce NRW, and perform pressure regulations and controls. The following is an outline of how non-revenue water is determined in District Meter Areas:

- Physical losses**:** because most DMAs do not contain any reservoirs or trunk mains, these components are typically not analyzed. Physical losses are typically pipe leakages on the main pipe, and service connections to the customers. This is usually higher at pipe joints. Analysis involves leakage monitoring at night, when the customer usage is low. The leakage flow is most likely constant throughout the day, while larger proportions of the flows are expected at night because the number of persons consuming the water is also low. The low flow of consumers at night diverts most of the flow to leakage flows.

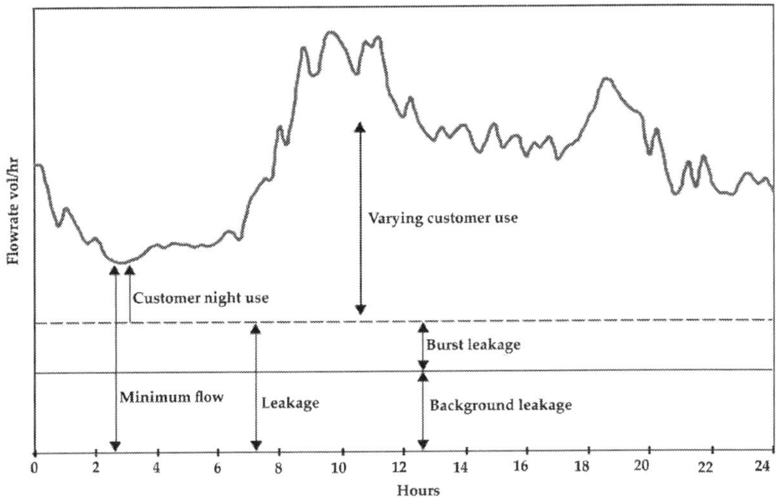

Source:https://www.researchgate.net/publication/51810756_In-pipe_water_quality_monitoring_in_water_supply_systems_under_steady_and_unsteady_state_flow_conditions_A_quantitative_assessment

Figure 8.13: Typical District Meter Areas flow profile

To estimate the background leakage, water utility companies analyses the night flow, taking into consideration the flow due to customer night use. The customer night use is termed the legitimate night flow. This level of flow can be estimated using analysis of the expected night usage based on the customer profiles of those served by that DMAs, then applying confidence limits. The night background leakage and the burst leakage are then determined by subtracting the estimated customer flows from the total night flows. This procedure is called the Burst and Background Estimate (BABE). This is represented by the equation:

$$Net\ night\ flow\ losses = Minimum\ night\ flow - Legitimate\ customer\ night\ use$$

- Commercial losses**:** these losses are determined by subtracting the night leakage flows (physical losses) from the total non-revenue water, which is determined by the total inflow less the amount of billed water in the DMAs. This is expressed using the equation:

$$Commercial\ water\ losses = Total\ nonrevenue\ water - Total\ night\ flow\ losses$$

$$= Total\ water\ input - Water\ billed$$

Once non-revenue water has been determined and detected in a District Meter Area, the process outlined before for non-revenue water reduction and management is used.

Monitoring Performance of Non-Revenue Water Management Strategies

Non-revenue water performance indicators are used to rate a utility's performance and operational efficiency against set standards. The major performance indicators used are:

***Expressing Non-Revenue Water as a Percentage of System Input Volume*:** this is the simplest indicator. It does not tell much about a utility's performance in its reduction strategy, but is useful as it provides an overall feel of the level of losses.

***Infrastructure leakage index (ILI)*:** this index measures how well the distribution network is maintained and repaired. It is an indicator for physical losses, and has no units:

$$ILI = \frac{Current\ annual\ volume\ of\ physical\ water\ losses}{Minimum\ achievable\ annual\ physical\ losses}$$

Other indicators look at non-revenue water as a percentage of the cost of running the system. To ensure accountability, water utility companies are required to establish independent auditing teams to measure the indicators, just like what they do to their financials.

9 THREAT TO POTABLE DRINKING WATER

INTRODUCTION

Water is one of the key natural resources that are of great significance to the survival of living species in this world. This is because it acts as a backbone to numerous aspects of life such as supporting lives, creating jobs and wealth, source of recreation and tourism. Therefore, if there is inadequate water supply, life in this planet would be threatened.

Statistically, about 97% of all water on the earth planet is not consumable (salty water) and only 3% is consumable (freshwater). It is also worth mentioning that about two third of these freshwaters represents the frozen water mainly in the ice caps and glaciers.

According to WHO, potable water can be defined as "*having acceptable quality in terms of its physical, chemical, and bacteriological parameters, so that it can be safely used for drinking and cooking*." In addition, potable water should be colourless, tasteless and odourless. It should be free from fluorine, arsenic, nitrates, and nitrite and lead contents. Also, potable water should not be associated with water borne disease during and after use. However, human activities comprising of both domestic and industrial discharges has resulted in the introduction of substances that is that are not supposed to be present in water.

Studies show that potable water is inaccessible in most parts of the world, especially in some of the developing countries. One major factor that has aggravated this problem is the fact that population growth has been increasing at an alarming rate, and this has made potable water inadequate. In other words, there is water supply problem in most places around the world which should be investigated.

In developing nations, disparities within metropolitan areas where access to water and sanitation services vary by geographical locations within the city and proximity to the pipe infrastructure, as well as household socioeconomic status is because of the rapid growth in urbanization. Urbanization can impede potable water supply and as a result, diminishing access to drinking water by households are greatly experienced. With the government not giving sufficient attention as to funding for the expansion and maintenance of the existing piped network, nature of settlement (unplanned) and inaccurate consumer profiling, efficient provision of water is compromised.

POTABLE WATER AND HEALTH

Water is one of the most valuable resources on the planet. Unfortunately, many places around the world is not only facing a drinking water shortage but also water contamination issues. Studies have shown that there is a very close association between access to adequate, clean, and safe water and health status of a population. According to a recent study that was conducted by WHO, environmental risk factors account for about 35% of the disease burden in children. Some of these environmental risk factors include unclean drinking water, lack of proper sanitary facilities, and improper hand washing and other hygiene practices.

Unlike many other environmental risk factors, studies have indicated that potable water (drinking water) is one of the key factors that can be altered, given the appropriate technology and adequate funding. World Health Organization (2010) studies indicated that about 1.5 million children that die every year are usually due to poor access to clean and safe drinking water. Lack of clean water is the leading contributor to diarrheal diseases in children, which accounts for 19% of the total child death worldwide. Therefore, in order to reduce the rate of infant mortality, the government have the mandate of ensuring the access to adequate, clean and safe drinking water for all. This is because research studies have shown that access to adequate, clean, and safe drinking water is more highly correlated with reduced child mortality compared to other socioeconomic indicators, (such as access to health care or percentage of households below the poverty line).

Cost-benefit analysis studies on potable water revealed that investing in water related project has positive returns and outcomes. For instance, World Health Organization (2004) found that a US$1 investment in water in the developing world yielded returns of US$5 to US$28. Most of the benefit consisted of time saved from gathering clean water or securing a private place to defecate, but it also included decreased morbidity and mortality and economic savings from avoiding diarrheal disease. This results to direct savind, as the need for health care decreases and indirect savings since, fewer people fall ill and are unable to work. It was estimated that providing water to each person in the world who is currently without it would cost US$22.6 billion, or US$10.7 per person in the developing world per year. This intervention would yield, an estimate of US$262.8 billion in economic benefits.

PUBLIC PERCEPTIONS OF POTABLE WATER QUALITY AND SAFETY

According to previous research studies that have been conducted before from different parts of the world, it can be concluded that the quality and safety of potable water for drinking is usually perceived differently based on locality and level of awareness of people. For instance, it is worth noting that education plays a crucial role in the way respondents perceive their potable water. This correlation may also be related to the economic situation associated with each method of receiving potable water. For instance, those with personal or household taps tend to have some level of education and live in more affluent areas. Those that live in areas of lower economic standard usually are good with communal taps. Communal taps are usually located within informal settlements, in which many citizens live in homes made of tin sheets or shanty houses. These taps are coin or card operated, while those people with household taps pay a monthly water bill.

Figure 9.1 below relates education with the perception of overall water quality. The same general trend was observed between those that university and non-university educated. Majority of them believe that the overall water quality is generally fair to good. However, more university educated believe it was just fair, while more non-university educated people feel it is good. Though reasons for these are not reflected in the graph, some university educated people, because of their level of education might have better awareness and information on quality of drinking water and its effects to health. Several studies elsewhere have been mentioning the correlation of education with respect to a person's ability to analyse and decide on matters concerning quality. As written in major textbooks, the safety of potable drinking water relies on its physical, chemical, biological and radiologic nature. Some non-university educated consumers usually just look the physical characteristics of water, indicating that "*if it is clean, tasteless and odourless, it is OK to drink*". This can also be the main reason why 8% of them believes that the water they drink is of excellent condition. As a conclusion, the university educated respondents are more particular of the quality of the water they drink, making them execute sound decisions on choosing its type of distribution their drinking water is to be and encourage themselves that paying water is more of an obligation and not an option.

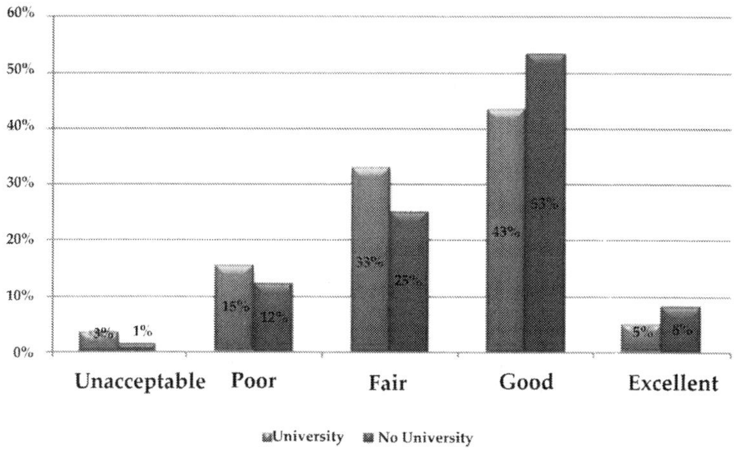

Source: https://www.ontario.ca/page/taking-care-your-drinking-water-guide-members-municipal-councils

Figure 9.1: Education vs. perception on overall water quality

Similarly, while most people consider the overall quality of potable water to be good, some people usually have varying views regarding the colour, taste and smell of the water. According to surveys, most people usually rank colour as the least desirable quality and smell as the best quality. In terms of colour, however, most people usually complain of the unpleasant colour, particularly during the rainy season when potable water often became a brown or cream in colour. Since this issue is not year-round, people usually reflect on their minor discontent by ranking the water as fair. Though improvements could be made to the water's quality, some people are usually satisfied with their potable water supply. Issues with taste are a concern primarily to those people with household taps, although several people in informal settlements may also face similar dissatisfaction. It is worth noting that those in the household tap communities may raise issues on distinct taste of over-chlorination. Regardless, the distinct chlorine taste is likely to be due to the chemicals used in the final stages of piped water treatment.

ENVIRONMENTALLY UNFRIENDLY THREATS TO POTABLE WATER

Water is hardly found in pure state, it contains a variety of contaminants. At times, relatively 'good' water is rendered 'bad' due to contamination. For example, spring water may be contaminated by decayed plant materials, bacteria, silt, clay, sewage etc. Borehole is contaminated by rusts from the water pipes etc.

Various activities being carried out within and around a place may lead to significant pollution and may possibly contaminate potable water. Among these activities include the intensive farming system. For instance, animal husbandry, fertilizer application, the spraying of insecticides and pesticides for pest control and bush burning by the farmers are potential sources of water contamination in most places around the world. The greatest threat to potable water pollution in an area around the world is sewage disposal, especially human waste disposal systems. Studies have indicated that most waste from toilets, urinals, laboratory sinks, and kitchens find their way into the sewage tank. After going through the treatment processes involving the partial breaking down of the organic wastes, the dry sludge from the drying bed and the effluents from the ponds are discharged to farm for irrigation and as manure. Poisonous gases like hydrogen sulphide and methane that bubble off from oxidation ponds also act as major pollutants of water (either directly or indirectly). Apart from sewage contamination on potable water within an area, the traditional habit of disposing faeces and urine in the bush and pits by rural dwellers is also pose threat to the quality of drinking water. It has been observed that most people in different places around the world depend solely on ground water supply. With substantial amount of water being drawn from the aquifer with powerful pumps on daily basis; the water level can be depleted thereby reducing its capacity to dilute potential pollutants and maintain proper salinity. Thus, with nearness of the boreholes to numerous environmentally unfriendly activities and high permeable characteristic of sandstone, the groundwater quality cannot be guaranteed.

One of the greatest fears in the use of potable water in households is the issue of its effect on health when the quality is compromised. Therefore, before using potable water for drinking, it is highly advisable to prevent the contaminants from being discharged to water courses that are currently used for drinking, rather than to purposefully increase the number available water. This approach is common in community development programs, where the residents have considered or currently practice water reclamation. Though pathogens are an integral part of water supply, it is the responsibility of water utility companies to take viable steps to prevent public exposure to these harmful elements.

According to a cholera study that was conducted in England on the fear of potential health complications from organisms in French and English polluted waters in 1875-1900, Escherichia coli, Salmonella, Giardia lamlia (pathogens that cause gastrointestinal disease) and viruses are some of the major threat of potable drinking water in most places. Many of these pathogens live or survive in water and infect humans who ingest contaminated water or foods, or through people-to-people contact. Studies led by the Water, Science, and Technology Board of the U.S. National Research Council have revealed that resources to make wastewater clean are available, however, there are uncertainties and potentially large risks associated with trace organisms and emerging contaminants.

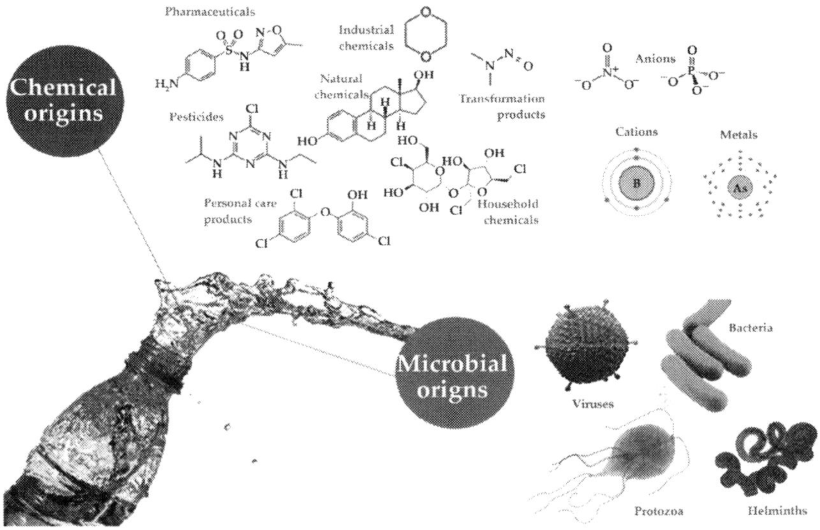

Source: https://www.wateronline.com/doc/drinking-water-protection-are-we-doing-it-right-0001

Figure 9.2: Key infectious agents in potable water

Some of potentially life-threatening infectious agents include bacteria, viruses, protozoa, and helminth. Before supplying water to household, utility companies should take additional steps to remove harmful chemicals present in potable water. To underscore the gravity of these life-threating agents in potable water, the following is a list of common infectious agents that may be present in potable water.

Bacterial Threat to Drinking Water

Bacteria are microscopic organisms that thrive under a variety of conditions. Ubiquitous in the environment, they live in a wide-range of living conditions. However, some of organisms are infectious, while some are harmless. Some are required for human survival, while others, such as Vibrio Cholera and Salmonella, are involved in causing the spread of diseases among humans. Salmonella and Vibrio Cholera are among the most common infectious agents found in drinking water supplies. Other common agents include Shigella and Helicobacter pylori. The most effective method of removing such bacteria from wastewater is by filtration.

Contamination at Treatment Plants and Piping Systems: Water treatment plant is one point where bacteria can be introduced into potable or drinking water. In other words, filtration at the treatment plants usually provide an easier way of obtained concentrated food supply for the bacteria to sustain themselves.

Source: http://www.merusonline.com/biofouling-refinery

Figure 9.3: Biofilm contamination of water piping system

Biofilm is a name given to a layer of biological material on the inside of pipes, tanks, etc. This material usually contains both mineral deposits and biological material. For reasons still not well understood, this material may, for certain periods, grow at an accelerated rate. As this occurs, some of this biological material may detach from the pipe, tank or treatment devices and be present in water samples. This material generally not seen by the unaided eye. It is difficult to fully kill this biological layer. If this material contains organisms of the total coliform group, these cells could lead to detection of total coliform. One cannot differentiate between biofilm bacteria generally present in pipes and bacteria originating directly from a source associated with disease.

Although biofilm is a possible cause, not all positive bacterial tests are the result of biofilm. It is critically important that all other technical explanations for the presence of bacteria be fully explored to identify the basic cause of bacteria in potable water. In these cases, wellhead samples should be taken after heavy precipitation to help differentiate well construction or soil filtration problems from bacteria associated with the distribution, plumbing or treatment systems.

Contamination of Dug Wells and Springs: Dug wells are usually about 3 feet in diameter and about 15 feet deep. Common construction problems with dug wells include a lack of mounded backfill around the outside of the well casing, insufficient casing height above the general ground level, inadequate or leaky well cover, and holes or unsealed joints in the sidewall of the upper portion of the well casing.

Source:http://www.seakayak.ws/kayak/kayak.nsf/a631f4790e66c273852569f9006ec1b2/3ebe21db81d65ba485256b32006a6835!OpenDocument

Figure 9.4: A typical dug well used for potable water supply in rural areas

Older wells made from fieldstone usually have many of these unsealed passageways and thus have very frequent bacterial problems. These well construction problems can sometimes be confirmed by looking for leakage on the inside of the well casing after a heavy rainstorm or simulated "rain" from a garden hose.

Construction problems in springs are like those of a dug well with one addition: the frequent entry of bacterial contaminants in the reverse direction through the overflow pipe.

Contamination of Bedrock Wells (also called artesian / drilled): Common construction problems with bedrock wells include a buried well head and an inadequate well cap. Installation of a pit-less adaptor should prevent the leakage of bacteria-laden surface water directly into the well. Construction problems with bedrock wells can sometimes be confirmed by identifying leakage on the inside of the well casing after a heavy rainstorm or from simulated rain from a garden hose.

Source: https://howlingpixel.com/wiki/Water_well

Figure 9.5: A typical bedrock well for potable water supply in rural areas

New Pump Installation or Plumbing Work: When a submersible pump is taken out of a well or when a new pump is prepared for installation, it is typically placed directly on the ground near the well. When this occurs, dirt laden with bacteria often adheres to the pump, the water discharge line, and/or the electrical power cable. This material then contaminates the well when the pump is installed. After reinstallation of the pump, time and flushing are necessary to remove this solid material from the assembly. Only after flushing to loosen this material should disinfection be conducted. When recent work has been done on the home's plumbing system, such as hot water tank replacement, pipe installation, etc., bacterial problems will likely be experienced for a few days thereafter. Very strong flushing, possibly followed by chlorination if necessary, will likely clean the plumbing system of bacteria.

Newly Constructed Wells: The installation of any new well normally allows substantial bacteria to enter the fractures of the bedrock or the soil around the outside of the dug well casing. Mud and soil particles protect these bacteria from disinfection. Sustained flushing is needed to remove this mud, pulverized rock and bacteria prior to disinfection. In rare cases, the removal of construction debris may take months. Please note that disinfectants cannot kill bacteria trapped deep within these construction residuals. The solids must first be flushed away before effective disinfection can be expected.

Viruses Threat to Drinking Water

According the World Health Organization (2010) report, viruses are cellular parasites made up of a protein coat and genetic material (DNA or RNA). They cause a variety of diseases, often discussed by the media. Some examples include Severe Acquired Respiratory Syndrome (SARS), Acquired Immunodeficiency Disease (AIDS) and influenza. Viruses are incapable of surviving outside a host cell, thus they invade and take over the metabolic machinery of other cells. Among the most common enteric viruses transmitted through contaminated potable water is Hepatitis A. Coagulation is the most effective method of removing viruses that commonly reside in contaminated potable water.

Protozoa Threat to Drinking Water

Protozoa are single-celled organisms, that are similar to animals in cell structure and nutritional requirement. They are mobile, that they move with the use of pseudopodia, cilia or flagella. These microorganisms typically live in water, as they require moist environments for survival. They commonly reside in ponds, streams, lakes, oceans, as well as moist soil, beach sand and decaying organic matter. Some protozoa reside in animal hosts (whom they originally infect), and in turn spread the disease further. These microorganisms are difficult to test since, in potable water, they co-exist as colonies living symbiotically with the other colonies. Common infectious agents, which threaten the safety of those who drink potable water is Cryptosporidium enteritis and Giardia Lambia. These two microorganisms are known to be resistant of the typical chlorination of potable drinking water. The common method of transmitting protozoa is through the drinking of contaminated water. Filtration is one of the most effective ways to remove protozoa from the water, while other modern treatments include UV, ozonation and metal hydroxide disinfection.

Helminth Threat to Drinking Water

Helminths are parasitic worms that cause a variety of diseases, particularly in areas with low sanitation. The common method of infection is through faecal-oral transmission of helminth eggs. About 12-90 percent of the eggs ingested develop into worms. Among the most effective methods of removing helminths from wastewater is through flocculation and sand filtration. Exposure to UV together with chlorination was also used to remove helminths from drinking water, then followed by filtration. Some studies have also proven the effectiveness of ultrasonication followed by coagulation and flocculation of the disintegrated material.

Endocrine Disruptors and Pharmaceuticals threat to Drinking Water

Endocrine Disrupting Compounds (EDCs) are chemicals that disrupt the function of the human endocrine (hormonal) systems. Scientists hypothesize that they cause cancer and damage multiple organs, including male and female reproductive organs and the thyroid. They pose a significant threat to the consumers of contaminated water. Common disruptors include dioxin, diethylstilboestrol (DES) and other pesticides. EDCs are synthetic chemicals that mimic or block hormones in the body upon consumption. These chemicals disrupt normal body operations by changing the function of the processes the hormones control. According to Natural Resources Defence Council, (1998), most of these chemicals usually cause hormone levels to change, altering the way they travel through the body, or by stopping or stimulating the production of hormones. Pharmaceuticals such as antibiotics, seizure medication, painkillers, and cholesterol-lowering compounds have also become an increasing threat to consumers of polluted potable water. These compounds, as well as preservatives and active ingredients in personal care products can cause effects similar to those caused by EDCs. Removal of these chemicals depends on their intrinsic chemical properties. For instance, chlorine is capable of removing 90 percent of certain compounds (those with aromatic carbon and amine or hydroxyl functional groups) and less than 20 percent of other compounds. Activated carbon is effective in removing over 75 percent of endocrine disruptors and PPCPs.

AWARENESS CREATION ABOUT THE QUALITY OF POTABLE WATER

Different research studies have shown that most people around the world usually have very little or no knowledge about the quality of potable water for drinking. Since education is crucial in order to understand the threat that potable water may be having in terms of different contaminants that are readily available in the environment, there are several possible strategies to better inform citizens about the process involved in supplying potable water in the hopes of increasing acceptance. Some of these recommendations include:

- Developing formal curriculum for primary schools
- Implementing potable water challenges among universities and campuses
- Using various media to distribute information about the importance of quality potable water
- Distribution of a follow-up survey

Development of a Formal Curriculum

To address the issue of young adults having little knowledge of the water system, the first action that could be taken is to work with the Ministry of Education to create a formal curriculum for primary schools that includes the topics of water supply systems and water conservation. The Ministry of Education could require that all primary schools revise the current curriculum to include these topics. By taking this step, a country can ensure that all future generations are aware of water resources supply systems and the importance of water conservation. Furthermore, children could share this information with their families, encouraging them to be more conscious of the water they use every day, to help them and motivate them preserve water for future water needs. Though not as easy as it seems, the effort must originate from the Ministry of Education, through a memorandum circular on curriculum revision. After these a group or panel of stakeholders composed representatives from the ministry of education, primary teachers, parents and students, as well as representatives from the community and child welfare be gathered to primarily draft the revised curriculum. Each representatives point of view must be considered in drafting the revision.

Implementing Potable Water Challenges among Universities and Campuses

Universities and campuses must also be included in addressing water issues. Since this group of consumers are in the higher degree of understanding with respect to the quality of potable drinking water, these group of people can be used to basically educate the citizenry on how they can maintain the quality of the potable drinking water.

Universities and colleges usually hold several events during the term where awareness to potable drinking water can be themed. Each university should include in their school calendar an activity or event in the campus or university whose theme is about potable drinking water. It can range from film showings, plays or events that highlights safe drinking water to science investigatory projects and researches targeting preservation of the quality of potable water during science week, or a week attributed to drinking water awareness. These projects can be even extended to the community level, to basically educated every age group of that community.

Last but not least, a university or campus providing potable drinking water to their students must have transparency and open access to students, parents or staff on the quality of drinking water they provide. Records of Potable drinking water test must be help at all times, as well records or issues and action taken to resolve issues must also be shown. In addition, the university must also encourage free and immediate reporting should deviations from potable drinking water quality is observed, as well as citing of breakage and leakages within the drinking water line.

Education through Media

A country can develop educational schemes using all forms of media. This includes newspaper ads and articles, radio announcements, billboards, as well as television commercials. Informational tours of a reclamation plant could also be offered two or three times a week, with one tour occurring on the weekend, so those citizens who work during weekdays can have the opportunity to learn about the process as well. One day of the week could be dedicated to school tours, so students can learn about the water supply processes and conservation in a hands-on way. The government can also continually print information in the local newspaper to ensure that all readers have the opportunity to learn about the all processes involved in potable water supply. It is also feasible to develop a radio segment that airs on all popular radio stations, given that most citizens listen to the radio.

Source: http://jasoninhollywood.blogspot.co.ke/2015/08/summer-drinks-and-lifestyle-billboards.html

Figure 9.6: An eye-catching billboard about saving potable water

To further educate the public, the Department could develop fun, eye-catching billboards that inform city members about free informational about potable water. To enhance the educational process, the Department could also distribute information with the local water bill; a newsletter would satisfy this requirement. This newsletter could be printed in the most commonly spoken languages and dialects.

Contact Number

The responsible department of infrastructure and Water Management can set up a calling system so citizens can call and ask questions about potable water related issues. This system could be partially automated. Pre-developed answers to frequently asked questions could be made available, and in the case that individuals have more questions further assistance could be provided by a potable water specialist. This line could also be made available for citizens to file complaints.

Follow-up Survey

Finally, the government can implement a follow-up survey within the next ten years to determine whether the educational campaign obtained the desired outcome. From those results it can then be determined if any additional actions are needed to be taken to further educate the citizens. To attain an accurate representation of the public opinion, the responsible department should calculate the number of surveys required. This number can be based upon the number of households within each suburb rather than the population in each area. Furthermore, surveys can be printed in all major dialects and languages in order to facilitate their rapid completion. Having the surveys in multiple languages would also ensure a more accurate representation of the population, as those unable to speak English will also be capable of answering the questions presented. Distributing such surveys with a monthly water bill, in household tap areas, and allowing respondents to return the survey via mail, would limit the amount of labour, and therefore the expenses incurred by the Department.

10 WATER CRISIS

INTRODUCTION

Water is one of fundamental natural resources important to sustain life on the earth. Erratically distributed across the globe, the transformation, processing and distribution of potable water to low access water points areas is one of the main problems in our world of today. Water contamination poses a serious threat to human health on a global scale. Paradoxically, even though our planet earth is 71 percent surrounded by water (96.5% of which is in the oceans), most of it is not safe for drinking. And providing safe drinking water for the world's population is a constant struggle. Today, especially in third world countries where almost 3/5 of the population do not have access to potable water, the situation is getting even worst.

According to the World Health Organization (2010), over 50% of the United States population depends on groundwater for drinking water. This water, highly susceptible to pollutants is mostly generated from; storage tanks, septic systems, uncontrolled hazardous wastes, landfills, atmospheric contaminants and chemical and road salts. Water scarcity or crisis can actually be increased in two ways (that is through natural water scarcity and through manmade water scarcity).

Source: https://www.worldvision.org.hk/en/learn/water

Figure 10.1: Water crisis is real

Poor people, particularly women, are the most affected in case of a water crisis in an area. They are often unable to engage in paid work when they do not have safe water nearby. This is because they often spend hours each day trekking to the nearest water source, waiting their turn in long queues for water, or is too ill with water-related diseases to have the strength to work. This has a negative impact on paid employment for the females or males that are employed since they are not able to constantly attend to employed work. The World Health Organization (WHO) estimates 5.6 billion working days would be gained annually if there was universal access to adequate, clean and safe water for domestic purposes. In contrast, people living near safe water supplies can look after the water needs of their family in a matter of minutes, leaving the rest of the day free to earn much-needed cash.

According to World Bank (2011), "*women being the main collectors of water it is often their lives that change the most dramatically.*" Thus, the scarcity of water leading to erratic water availability implies that females spend more time in water collection activities. If the water supply issue is addressed in many countries, it would be available closer to their homes; people will have additional time freed for other activities and decision making.

There are certain factors that are aggravated by these two types of water scarcity. For instance, studies have shown that hydrological change is one factor that may significantly be affected by water crisis in a region. Other factors include climate change and water management issues among other factors.

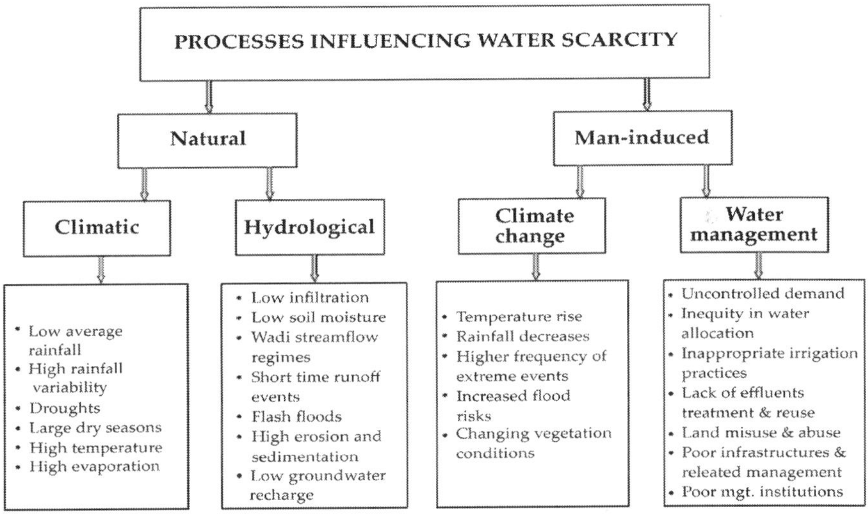

Source: http://www.sciencedirect.com/science/article/pii/S0022169415003388

Figure 10.2: Illustrating natural and man-induced processes favouring water scarcity

According to Water Aid (2001), access to water impacts the psyche of a community in several ways. In communities where access is limited and people (water collectors) must walk long distances to fetch water, there is increased tension from the safe arrival of the female family member that fetches the water. This is because some water points are in disserted areas that are unsecure. There is also reduced self-respect since most of the attire they put on as they fetch water gets dirty.

Despite the many negative impacts, there is also the positive aspect in it, such as; employment opportunities for personnel employed at the water distribution points. They earn a living out of that; also, the water vendors who deliver water to consumers on payment hence a source of livelihood; water kiosk owners, they distribute water to consumers on payment thus make a living out of that.

CURRENT WATER USAGE AND REQUIREMENTS IN THE WORLD

***Water Usage*:** Currently, 35% of human water use is unsustainable, drawing on diminishing aquifers and reducing the flows of major rivers: this percentage is likely to increase if climate change worsens, populations increase, aquifers become progressively depleted and supplies become polluted and unsanitary. From 1961 to 2001 water demand doubled - agricultural use increased by 75%, industrial use by more than 200%, and domestic use more than 400%. Humans currently use 40-50% of the globally available freshwater in the approximate proportion of 70% for agriculture, 20% for industry, and 10% for domestic purposes and the total volume is progressively increasing. Current percentages for worldwide fresh water usage are as follows:

- 70% of all the fresh water is used for agriculture - Countries are importing food as a way to import water. Producing one ton of grain requires 1000 tons of water, but producing one ton of beef or cotton requires 15,000 tons of water.
- 20% of Fresh Water is used by Industry - As water becomes scarce, demand for water in cities and by industry is satisfied by taking water from a country's agriculture.
- 10% of fresh water is used for domestic use - Domestic use accounts for three-fourths of the urban water demand. For instance, each day in the U.S., more than 4.8 billion gallons of drinking water is flushed down to the toilets.

Water Requirements: The main consumers of drinking water are private households, small commercial operations and industry. Drinking water consumption for nonresidential buildings is heavily dependent on use. Hotels, hospitals and senior care homes have a proportionally higher requirement that can be accounted for, primarily, through washing and catering needs.

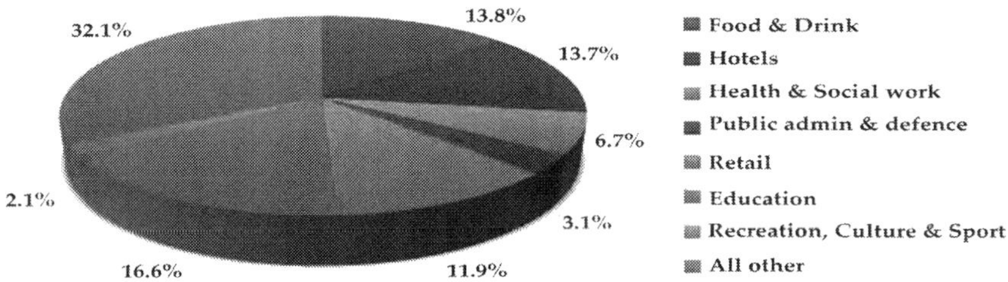

Source: http://www.zanran.com/q/public_water_supply_england_and_wales

Figure 10.3: An example for water usage distribution: main commercial/industrial sectors in England and Wales

According to the above figure, it shows that hotels are in the average zone of water use rates exceeding those of health, social, retail, recreation, culture, sport and others. The sustainable use of water implies resource conservation, environmental friendliness, technological appropriateness, economic viability, and social acceptability of development issues. The perception that water is increasingly scarce gives to water management a great relevance.

CAUSES OF ERRATIC WATER SUPPLY

The possible causes of erratic water supply are usually varied depending on the governance and management systems of a place.

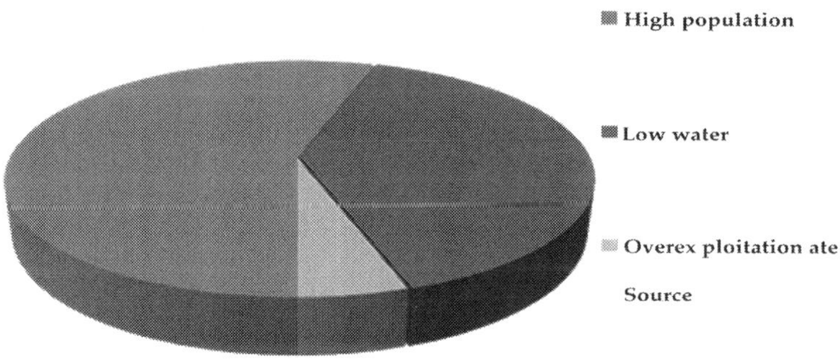

Source: https://www.slideshare.net/OlusanyaOluwatobi/accessibility-to-water-supply-and-research

Figure 10.4: Causes of Erratic Water Supply

Figure 10.4 illustrates the 3 major causes of erratic water supply. It was revealed that high population is primary cause of erratic water supply. The high population in a specific are produces a high demand on water supply which magnifies the source of the scarcity. As the population continues to climb up, the water supply becomes inadequate to a point that re daily water allotment per person in a community might not be achieved. The second major reason of erratic water supply is the low water flow or low water pressure. This is mainly attributed to the present capability of the facility to supply water in a specific community. Good water supply in an area is achieved when water flows as expected. Water flow is important since this determined how fast or how long a person can perform the activity of using water. However, water flow become sluggish specially when leaks are present, and it worsen when leaks becomes undetected. Due to this, at times the water force is very minimal for the water to move up the system translating to no water in the pipes and taps until the pressure is higher when the water level goes up. This may take longer especially during the dry seasons. According to studies, low water pressure is a main cause of erratic water supply since the water has very little force to move up the water table and higher distances resulting to prolonged lack of water in the system and may require electricity to pump up the water. It is expensive to install pumps and pay for electricity.

The last major cause of erratic water supply based on Figure 10.4 is the overexploitation of water at source (water catchment points). For instance, in arid and semi-arid regions, the amount of rainfall has no definite patterns, and that the problem in water arises when no rain at all comes for a couple of weeks or months. In some places, the summer season has the capacity to dry out the surface water, leaving people no choice of using ground water as a replacement. As such, boreholes and dug wells becomes a very interesting structure to be set near your home. The numerous boreholes and dug wells being constructed depletes the ground water aquifer. When the recharge of the ground water is becoming too slow with continuous withdrawal of water, the source groundwater later dries out resulting to water shortage.

Reduced water levels at the sources are also reported as another cause of erratic water supply. The water source points e.g., boreholes and wells, have been degraded in most places around the world due to lack of proper management and conservation and over exploitation leading to low levels of water available for consumption. Pollution at the source points is also a cause of reduced water available for consumption. Other minor causes of erratic water supply include overexploitation and power rationing or frequent power cuts.

SOCIO-ECONOMIC IMPACTS OF ERRATIC WATER SUPPLY

Waterborne Diseases

There are numerous water borne diseases that may arise because of erratic or unreliable water supply in an area. Some of the identifiable diseases include diarrhea and cholera and their effect on the households are usually categorized in terms of employed, business persons and the unemployed. The results obtained were represented in the table below.

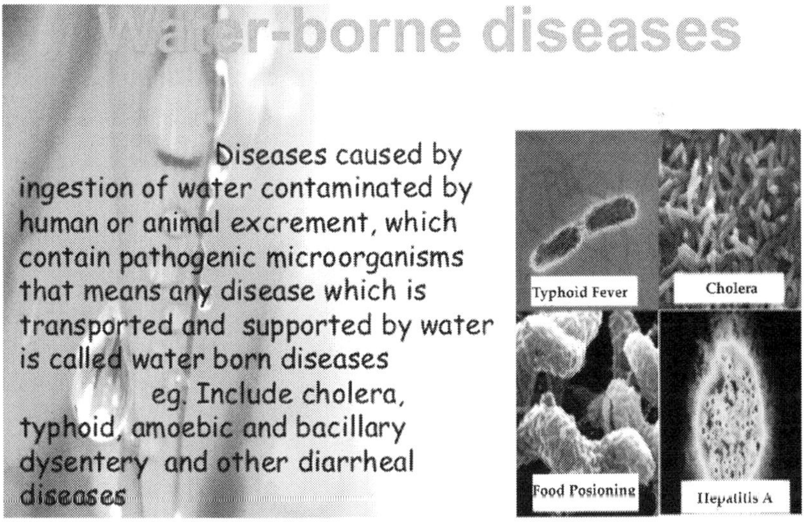

Source: http://preparednessadvice.com/medical/common-water-borne-diseases-and-how-to-treat-them-without-modern-medicine/

Figure 10.5: Water-borne diseases

Due to reduced quantities of drinking water, some people are usually forced to take chances. They buy and consume water which they are not sure of the sources and at times are not fit for consumption leading to the above-mentioned water borne diseases (diarrhea and cholera). This usually impact negatively on the health of some members of the population resulting to low output in terms of labour. They may not perform their duties as expected hence, resulting in a reduced output in production which negatively affects the economy and development of a community and of the country at large. The water borne diseases require treatment which requires finances. Those finances would have been used on other activities that would generate income.

There are other minor social-economic challenges that some people face due erratic water supply. These challenges include physical damages related to collection of water. Some of these physical damages include cases of spine damage due to water collection and cases of hip damage due to water collection.

Most of these physical damages are due to water collection methods such as carrying water containers on the back which negatively affects the hip joints or on the head which affects the spinal cord due to the heavy weight of water. The physical damage may also be attributed to the distances travelled when transporting water. If one is used to carrying water up the stairs, up a hilly road or for long distances, it may negatively produce a significant effect on the muscles and bones. If this happens a lot of money could be spent to treat the patient, and this would negatively impact on the economy of a place. The physical damage may also be permanent which translates to low output by the individual which affects the economy of a place.

Waste water drainage systems are blocked by solid waste that requires lots of water to push down the waste. The communal toilets require a lot of water to flush down the waste but due to erratic water supply the water is not adequate to maintain the right standards of hygiene. Communal washing sinks require a lot of water to be clean, but due to erratic water supply in a place, it is not always available for cleaning and most people are forced to depend mostly on re-used water.

Water Related Conflicts

Water related conflicts are key social-economic effects that are likely to arise in an area due to erratic water supply. There are some issues that are likely to cause conflict such as water shortage, overcrowding at collection points, limited water supply, and prolonged queuing time. In most places around the world, conflicts related to water are very common due to the demand for the scarce resource. Being a basic resource, every individual, struggles to obtain the resource leading to misunderstandings which result to fighting at times. The erratic water supply leads to fights since the water is only provided for few hours/weeks in most places, especially in urban areas.

Reduced water sources in an area usually result to limited water collection points. This leads to queuing for water especially the few hours of the day(s) when there is running water at the pipes. More than 75% of the people, especially in developing countries usually spend much of their time fetching water for domestic uses.

Source: http://www.sswm.info/es/content/decentralised-supply

Figure 10.6: Women waiting to fetch water from a common point

The figure above indicates some females waiting to fetch water from a common collection point. Females are the frequently involved in gender in water collection and are the most affected when there is erratic water supply in an area. They spend a lot of time waiting to fetch water which usually results to time wastage especially when the queues are long. In queuing for the water, they spend a lot of time which would have been used in other developmental activities.

The figure above also indicates that there is a significant association between queuing at the water source and the gender. The females who that queue are higher compared to the males since the females are the most affected gender during water collection in most households.

Poor Sanitation

According to World Health Organization (2010), it has been revealed that there is a close association of poor sanitation/low standards of cleanliness and erratic water supply in an area. This is in line with what was revealed on water shortage by the UNICEF (2011) that water has a major effect on the sanitation of an area especially in the urban areas.

WATER SCARCITY IN DEVELOPING COUNTRIES

Water scarcity is one of the most misunderstood issues within the globe but is one of the most threatening issues in developing countries such as Asia and Africa. Less developed countries are mostly located in tropical and subtropical landmarks where the industrializations of some of these countries are determinant to access to water. There are numerous research studies that have been conducted that states that the importance of countries, specifically African countries, to develop appropriate systems and programs to combat water scarcity based on the rapid growth of the population. According to World Bank (2010), population growths usually result to increase in demand of water and this cannot suffice the needs per number of people within a country. When a country reaches or exceeds 1000 persons per million cubic meters of water, the absolute water scarcity is usually reached. According to research done by the World Resources Institute (1986), it has been revealed that Africa populations are in the midst of quadrupling since 1986.

It is also worth noting that access to water is a necessary condition for development, which becomes relative to most developing countries. For instance, urban areas in developing countries tend to have a higher water productivity compared to rural areas. It has also been proven that the urban areas tend to have priorities in distribution and reallocations in water as compared to rural areas which have low priority for distributors to the agriculture. The reallocation of water stems to be high in costs and environmentally-urban cities' water transfer generally result in more wastewater being generated which in hindsight raises the risks of contaminants and pollutants within the industrial areas. Accessing water can be affected based on the distance of the water sources from a household. For instance, if people have to walk an approximate distance of 2 miles to collect water for 30min and ambulate back to utilize the water for domestic purposes; adequate hygiene will still not be achieved. The frequency of the access to the water source is not enough to produce enough water that can be sanitized and utilized for hygienic purposes.

The potential of water scarcity to become an issue based on population growth and lack of resources to develop water programs that are sufficient is integral to the rural areas in developing countries because of the lack of economic resources and irrigation infrastructures available to provide quality water for domestic and agricultural usages. Water scarcity potentially has affected migration patterns and other demographics such as death rates in developing countries. Therefore, creating awareness of the implication in water scarcity by political leaders, organizations and planners in developing countries is greatly necessary.

DEALING WITH WATER SCARCITY IN DEVELOPING COUNTRIES

Most countries around the world do not consider the value of water and are already causing various water resources (such as rivers) to run dry before they even reach the sea. It is also worth noting that most water utility companies usually charge for the delivery of the water and charge their customers by volume of water used. Current water policies are inadequate for addressing the needs of seven billion people at current consumption levels. As our global population continues to grow, maintaining current rates of consumption of existing freshwater supplies will likely lead to freshwater shortages.

Conservation of water is one of the critical strategies that need to be embraced as freshwater becomes scarcer. Water conservation measures are actions, behavioural changes, devices, technologies, or improved designs or processes implemented to reduce water loss, waste, or use. As the world's population continues to grow, people will continue to migrate to large urbanized cities. A cities ability to provide freshwater will be challenged as its population grows. To address these challenges, cities should typically try to gain access to more fresh water by purchasing water rights, building dams, etc. However, there is another way to address the needs of the population: this is through conservation. Leaks in a home can contribute to not only a large waste in water but a lot of money that the consumer loses for water never used. According to a research study of 1,188 homes in 12 cities across the United States, 67% of homes leaked an average of 10 gallons per day or less. But of those 67%, 5.5% of homes leaked an average of 100 gallons or more per day, which accounted for 58% of all water leaked. This illustrates how much water can be lost or saved by resolving issues on an individual level.

There are other issues that may arise in meeting the needs of water consumers. For example, the cost of water isn't the true value of water. Water utilities charge consumers for the delivery of the water they use. "*Although reduced residential water use and "water conservation" are generally portrayed as positive in water-poor regions challenged by restricted water supplies, the observed gradual attrition in residential consumption may force utilities to raise rates to cover basic operating costs and maintain sufficient revenues for expanding service and replacing old water mains and equipment*". As conversation continues and climate change continues to affect the globe, water prices will continue to rise and reach their true cost which includes the cost of not only delivering the water, but the cost of gaining the rights to the water, the environmental impact of diverting the water from other areas, ensuring water quality, etc, as well. The true cost of water could end up being a price that is not affordable for everyone, which could lead to regulations ensuring that water remains affordable for all demographics.

Awareness: When individuals are educated on water scarcity issues and water saving techniques; they will make choices that will lower consumption at the individual level. Creating awareness or sensitizing people about the importance of water security is one way to deal with various challenges of water experienced in different places, especially in developing countries. One way of creating awareness among people involves educating them on various alternative ways to produce water for their respective homes based on the inconsistent availability of water in public water sources and the effects it has on their living conditions. According to WHO (2010), "*sometimes water might not be available in households for up to a week. In some places people are forced to go outside of their homes in order to retrieve water for their homes. Sometimes, people are even forced to spend a lot of money to buy water for their domestic usage, and those without money it becomes highly impossible to get this previous commodity.*"

Though awareness seemed to be a common factor amongst different people in most places around the world, studies have revealed that there is a lack of awareness about water scarcity relief efforts in developing countries. For instance, majority of the populations being in more urbanized areas lack awareness or exposure to various water relief efforts provided by NGOs and CBOs. This is because most of these non-governmental organizations are usually located in the rural areas.

One intriguing dynamic view is that despite the fact that there are established awareness programs of water scarcity and even a mutual experience with water scarcity, there is still inadequate initiatives or effects taken on how people from different places, especially in developing countries can utilize water in order to better their living conditions.

However, it can be said that gender plays a certain role in awareness creation. For instance, most females are usually cautious when it comes to water management in households because they are the one involved in the water collection duties. This means that women can be used to create awareness (whether direct or indirect) protocols among people in order to deal with water crises, especially in developing countries.

Public education: Future efforts to encourage water conservation should employ public education and campaigns to gain the support of the community.

According to past researches, (UNICEF, 2008) environmental knowledge and values explain nearly half of an individual's likelihood to conserve. Public education is one of the highest rated policy options that can be used to garner community support before pursuing conservation measures in terms of water security issues. Public education campaigns should be led by multiple stakeholders to reach different parts of the population and to increase their legitimacy by incorporating multiple sources of authority. Because of generally low involvement in local conservation, these efforts should engage local organizations that are not directly related to environmental topics, such as church or volunteer groups. Based on past literature, public education and campaigns are most successful when they frame issues in a local, rather than global context and when they decrease the perceived costs and increase the perceived benefits of a policy.

These past findings about effective forms of public education are supported by results of recent studies conducted by UNICEF (2010), in which local water concern was significantly associated with the decision to conserve water. Planners and utilities can work together to communicate that the region's current water usage is unsustainable and to propose methods by which the community can avoid a looming shortage. Studies have also shown that environment has the most important influence on the decision to conserve water.

In other words, all planners can increase public support of conservation initiatives by emphasizing the ways that residents can benefit from their local water resources just by shifting their behaviour. As a first step, education campaigns could provide information to local builders and homeowners about the EPA's Water Sense program. Planners might also be able to cooperate with local utilities to provide a subsidy for water-efficient products and encourage their implementation.

Past research has demonstrated that public education alone is not likely to produce substantial long-term reductions in water demand. Given this knowledge, planners need to explore regulatory options for reducing water demand. Survey responses indicated that most people are usually concern about their natural water resources and motivated to protect them, but these factors do not always translate into conservation behaviour. Built form regulations, it may have a long-term impact on local water use and complement individual decisions to conserve water. Planners should rely on extensive community involvement throughout this process to solicit their input and build support for their implementation.

Creating Concern: Although awareness is high amongst most people around the world about water scarcity, an emerging factor that should be included in any program tailored toward dealing with water crises in a place is creating concern about water security. Based on a research survey that was conducted in Kenya to find out the level of concern among different people with water issues, it was found out that many of the participants didn't seem to have water scarcity as a high priority of concern in their daily activities. Even though being acknowledged as a prevalent issue in their native lands, the common consensus of the participants indicated that their main concern in Kenya is the utilization of electricity. It was repeatedly mentioned that the solution to water scarcity has a direct relationship with solving the electrical issues in Kenya. This was a statement given by one respondent based on that survey; "*when there's no light, there's no water for you because based on the borehole provisions, you have to use light to generate the water. So, you know so basically that you just try to save the little you get when there's light.*"

This statement from one of the participants captures the new emerging factor of concern about water scarcity. In other words, it is clear that water scarcity seems to be a subcategory that lies within the electrical utilization in a place, therefore lowering concern of the effects that water scarcity can have in different homelands, regardless of how high awareness of water scarcity is. Therefore, concern is one of the key determinants in the relationship between the awareness of water scarcity and the contribution to water scarcity.

Integrated Water Resources Management (IWRM): The WHO (2005), stated that water security is defined as "*the availability of an acceptable quantity and quality of water for health, livelihoods, ecosystems and production, coupled with an acceptable level of water-related risks to people, environments and economies.*" Reduced water security, or scarcity, is not just a measurement of decreased precipitation, but also considers factors such as the relationship between supply and demand over time and across a region, individual or community access to water, changing climatic conditions or water demand patterns, and livelihood and resiliency to cope with scarcity. Drought is simply one component of water scarcity.

Water resource management "*commonly seeks to reduce natural variation in target resources because fluctuations impose problems for the industry that depend on those resources*". Physical processes, no matter how engineered, are still highly variable and unpredictable, especially in the prospect of climate change and climate variability. Water resource management, although it seeks to stabilize water supply, may also contribute to perceptions and experiences of water insecurity or shortage.

The WHO (2011) suggests that water management activity should be classified as integrated water resources management activity. Based on this framework, there are certain elements that should be attained in order to have water security within an area. These include:

- Allocating the framework within the largest possible hydrological unit (when doing this, minimal management should consider upstream-downstream interactions
- Taking the complete water cycle into consideration even those diversion brought into the cycle by human activity (sewage and storm water)
- Focusing on sustainable development

Integrated Water Resources Management as a water resource management framework is important because it provides several solutions in terms of the general water resource management challenges that might be experienced within a place. Importantly to this however, is to actually understand the entire Integrated Water Resources Management framework in order to enhance the implementation process of the framework. Therefore, responsible water resource managers should have knowledge about or have access to information about water resources in order to effectively and successfully undertake the implementation of the framework.

It is worth noting that the prevalent opinion seems to be that water is scarce, is probably premised on a lack of access, socially-constructed scarcity and marginalization emanating from the apartheid policy making era. In other words, the perceptions of water scarcity in an area could not be refuted despite the availability of other economic activities such as irrigation. Using available gaps that may exist between land and water availability as the main arguing point, the national government in collaboration with various farmers within a country should work together in order to assist in dealing with water crisis.

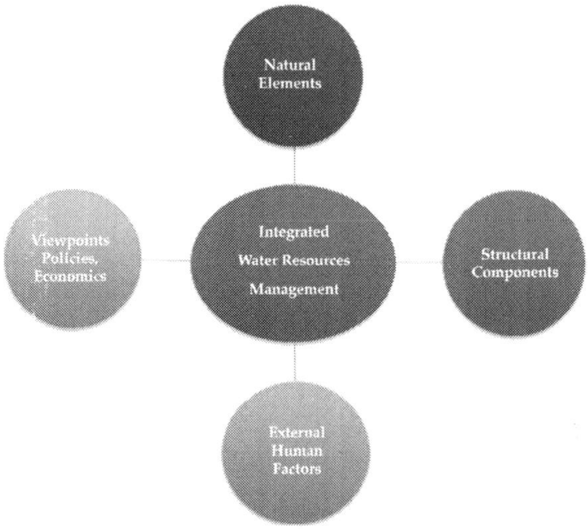

Source: http://integratedwater.blogspot.co.ke/p/introduction.html

Figure 10.7: A typical framework of Integrated Water Resources Management

There are several strategies that can be employed with regard to dealing with water shortage in an area such as improving irrigation efficiencies; valuing water as and economic good but at the same time offsetting this principle with well-targeted government subsidies to meet the equity requirements; increasing the number of households engaged in commercial agriculture, thereby expanding the beneficiary base and the per capita subsidy enabling the purchasing of a larger portion of the irrigation rights from the responsible Water User Association; and trading water with other economic development sectors, such as tourism could possibly expand the farmer's perspectives of water scarcity.

A vital step in the Integrated Water Resources Management framework implementation process would be to develop an integrated water resources management plan. Water Services Development Plans in different countries around the world lack the insight and inputs from both a water users and managers perspective.

Inclusion: Another very important strategy of dealing with water issues in a place is to include all responsible partisans in issues to do with decision making process that would influence how various water resources are managed, in both at national and community level. The issue of inclusion is considered as a crucial building block for a process-based approach to water security that is communicative and cosmopolitan. This is because it usually allows for a radical opening of the normative space of security by blurring the distinctions between insider and outsider, citizen and non-citizen, self and other. It also promotes a politics of recognition whereby all parties affected by water scarcity are provided with fairness and political opportunity.

There are two key ways in which inclusion work in respect to the context of water security can be accomlished. First, at the individual level, they break down essentialized views of identity, so often the cause of entrenched, intractable, and antagonistic differences. Secondly, at an international level they blur the borders of inside/outside, that fundamental characteristic of the Westphalian view. Together, both ways work to ensure inclusion is a concept necessary to overcome the essential "otherness" emblematic of conventional approaches to water security. An emancipatory water security relies upon a constitutive-relational theory of identity.

In conjunction with inclusion, there must be dialogues or communications when making decisions that are likely to affect the issues that deals with water security within an area. In other words, all voices within a discussion over shared water entails should be included or involved in order to promote the greatest possible freedom for individuals. Such a commitment draws from a deeper epistemological understanding that the mind plays a central role in the construction of reality.

Adaptations of Conservation Strategies: There are a number of goals that should be emphasized with regard to conservation of water resources within an area. These goals include:

Sustainability: water resources sustainability is of paramount importance especially in ensuring that there is adequate water available for the generations in future. Therefore, all withdrawal points for the freshwater from the entire ecosystem must not go beyond the natural rate of water replacement.

Energy Conservation: significant amount of energy is used in different places to pump, to deliver, and also for waste water treatment. In some places around the world over 20% of the total electricity consumption is devoted to water management. Therefore, for efficient water replacement an equivalent amount of electricity corresponding to the addition of water resource facilities must be achieved. In short as the number of water resource facilities in an area in increased, a more efficient manner of electricity conservation must be exercised, or if possible, an additional renewable energy source must be utilized to support additional water resource facilities for energy efficiency and economy.

Domestic Conservation: there are several steps that can be taken at community level in order to conserve water in various households. For instance, people should be sensitized on different ways of minimizing water usage and wastage in their households. Households must be first properly educated of the benefits and the technicalities of conserving water. Then a motivation to adopt conservation methods within their homes must be implemented to make such conservation effort as collective effort. The awareness must therefore, be installed in all the members of the household, and not just women members of the house. For instance, a kid can be taught how to conserve water by sing a glass when brushing their teeth instead of running water from the faucet. In return, they must first see this among adult member in the household.

Apart from the above outlined approaches or strategies of water conservation, there are other numerous approaches that can be employed in order to counter various water challenges in a place. For instance, there is the need to practice flexibility when making water related decisions. This is in particular to incorporating novel information, strategies, and priorities in water use and management. The main adaptation to erratic water supply that people should consider is conserving the already available water. This can be done indoors and also outdoors. In indoors, the bathroom is where you can save most water. This involve toilets (water closet toilets that require flushing) and cemented floored latrines that require daily washing, bathing and showering, basins and sinks. Toilets use a lot of water to get the waste away.

When hand washing utensils, it is important to use water sparingly. Close tap when washing dishes to avoid water wastage and rinse utensils in a basin preferably. There are other consideration and practices that have been proposed with regard to reduction of domestic water usage. These include; use of bucket or basin during cleaning instead of using running water, scrubbing floors on wash days to enhance water re-use, using sponge mops instead of rag mops since they require less water for mopping and they clean quickly.

Adaptations to erratic water supply also involve outdoor conservation measures. According to WHO (2010), some of the main conservation measures include:

- Re-using of water used to cool off and in wading pools to water flower beds, kitchen gardens, shrubs and lawns
- Keeping watering schedule flexible. For instance, after heavy rain, most people usually do not water lawns until it needs to be
- Interruption of automatic sprinkler systems, after raining if necessary
- Checking outside taps daily to see that they are turned off when not in use to avoid water wastage
- Checking shut-off valves for leaks often to reduced unnoticeable leakage and making sure all faucets are turned off completely each time they are used
- Replacing washers on dripping faucets and showers to reduce the amount of water to be used
- Replacing damaged or leaking sprinkler heads on underground sprinkling system to avoid water loss

All people with swimming pools may adapt to erratic water supply by: filling the pools with the right level of water to control spill-overs that may occur; covering the pool when not in use to prevent evaporation, minimize replenishment, and keep it cleaner; do not drain the pool unless repairs are needed to prevent water wastage. Erratic water adaptation strategies also involve the use of storage containers to store water to ensure availability when needed for use. Most people globally have resulted to storing water because of its scarce nature.

WATER RIGHTS

The right to water includes the right to clean and accessible water. In some places around the world, different people have varying problems related to access to clean and safe drinking water. Some usually resort to drinking dirty water from hand dug wells and other unprotected water sources. The picture below shows an example of a well, with very dirty and little water, on which people depend on as a key source of water for both domestic water and water for livestock.

Source: http://www.voyagerloin.com/15-faits-que-vous-devez-savoir-a-propos-de-l-eau-le-besoin-vital-73.html

Figure 10.8: Illustrating the situation of water quality in some places around the world

The practice of digging wells to cover the gap where the government fails to fulfill its obligations is indicative of the social adaptive capacity of the affected communities. The picture above is an expression of customary rights to water and the social adaptive capacity of rural communities.

It is not surprising that most people around the world usually resort to dirty water especially in times of lack of water at water points created by the government. This can be regarded as an exercise of customary water rights. According to UNICEF (2010), customary water rights are frequently rooted in customary land law, i.e. the body of rules and practices which govern access to and tenure of land. The UNICEF examines the situation in Nigeria, where a customary grant of land generally confers rights on water resources, among other products of the soil. In some places around the world, customary water rights have been implied by the courts in aboriginal title to land, in treaty-based land rights, in land reserve rights, and in common law land-based riparian rights.

According to World Health Organization (2010), it has been shown that lack of access to safe water has a major effect on people's health. Furthermore, poor health constrains development and poverty alleviation. Poor water and sanitation have an impact on education, but when safe water and appropriate sanitation are provided in schools, increased attendance and a reduction in drop-out rates results.

Based the argument on world standards, it is important to state that "The right to water clearly falls within the category of guarantees essential for securing an adequate standard of living, particularly since it is one of the most fundamental conditions for survival."

THEORETICAL APPROACHES TOWARDS HAVING REFORMS IN WATER RESOURCES MANAGEMENT

According to researches, it has been proven that protectionist approach is involved in creating a range of social conflicts endangering the future of natural water resources. This theoretical approach seeks to redress the injustices of the past by considering the needs of the locals and allowing them to participate in and to derive benefits from the concept of conservation. This new counter narrative is centered on the theme that "*conservation will either contribute to solving the problems of poor people who live day to day depending on the natural water resources or those resources will disappear*".

In the last few decades, the shift from fortress conservation to a more 'people centered' approach has been well documented and has spawned an overabundance of policies and programs. Some of these programs include; Community based conservation, Community water resource management, Community-based natural resource management, Collaborative or joint management, and Integrated conservation and development programs among others.

These people-centered approaches may vary greatly in level of participation, and can be analysed using Arnstein and Sherry's ladder of citizen participation in decision making processes.

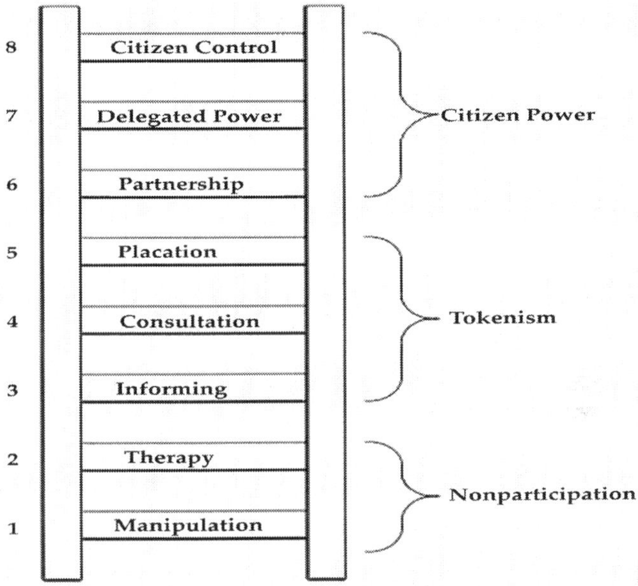

Source: http://www.errc.org/en-research-and-advocacy-roma-details.php?article_id=4174&page=0

Figure 10.9: Arnstein and Sherry's ladder of citizen participation in decision making processes

At the bottom of the ladder represents the local people, who are usually ignored when water resource management decisions are being drafted by a central government authority. At the top of the ladder represents a community-based water conservation approach, where local people establish and control an area of land and make decisions regarding the use of water resources within the area. In between the two poles are various arrangements between local people and central governments with varying amounts of local participation (such as joint management and other programs initiatives).

In some parts of the world, community water resource conservation has taken the form of Community Based Natural Resource Management. It involves the devolution of management responsibility to local people, the notion of water resource utilization rather than preservation, and the idea that markets should play a role in shaping incentives for conservation.

There are two separate themes upon which community-based programs can be based. The first is the "conservation and development" argument, which arose with the first Integrated Conservation and Development Programs, whereby benefits and incentives ('development') are provided to local people in the hope that they will be encouraged to adopt conservation-friendly attitudes and behaviours.

Experience with past benefit distribution programs has indicated that benefits alone may not encourage water resource conservation attitudes and behaviours. For this reason, therefore, community-based natural resource management should be based on a second theme. The second theme is centered on devolution of control over water resources, the 'community control equals conservation'. In other words, community-based natural resource management should seek to link conservation and rural development by drawing heavily on the participatory methodologies of modern development theory. More specifically, community-based natural resource management should aim to redistribute social and political power by devolving management and decision making to communities and allowing them to benefit from the management of any natural water resource. The underlying premise of this theme is that if communities are given sufficient authority and control over water resources, they can benefit from their management. Put together, these two themes indicate that benefits may influence conservation attitudes and behaviours if there is devolution of management and community control over water resources.

WATER SUSTAINABILITY

According to United Nations (UN) views, sustainability can be defined as "*development that meets the needs of the present without compromising the ability of future generations to meet their own needs*." In order to understand what water sustainability involves, there are certain approaches/principles that can be applied to explain it. These principles include blueprint for green economy principle, Economic principles, the user-pays principle, and the payback principle among others.

According to the principle of blueprint for a green economy, the real price for water, water management and water pollution control coupled with the cost to those caught in urban poverty, implies a socio-economic problem. Fundamental to an understanding of sustainable development is the fact that the economy is not separate from the environment in which we live. There is interdependence because the way people manage the economy impacts on the environment, and because environmental quality impacts on the performance of the economy.

Based on this argument on environmental economics, two general classes of environmental problem usually arise. First is the issue of water pollution: It is the introduction of materials not inherent to the water in the environment, and is essentially the residual of human activity that adversely affects the next user of water resource, and even the resource itself, and therefore man's well-being.

Secondly, there is the issue of water resource destruction. In other words, this is the removal or dispersion of water resource concern of materials that impairs some ecological process or depletes water resource otherwise capable of yielding benefits in perpetuity.

Water resource destruction and pollution problems both result from the reallocation of resources through economic activity, of which there are three, namely, land (all-natural systems in the biosphere), labour and capital. Economics is the science of allocating scarce resources among competing ends.

According to the economic principles, together with changes of habit in the more responsible consumption of resources, appropriate technology is at the core of water resource conservation solutions. But, the most effective mechanism for checking the culture of rampant consumerism lies in the application of economic principles.

Based on the user-pays principle, it is constructive to conduct a research program aimed at educating the consumer into the social costs of their use of water resources. A captive user would, for example, be a visitor to a nature reserve. They could have the option of being supplied through metered water resource-saving devices that have been installed in the units. User-pays measures have the potential to significantly influence the pattern of use of key resources through economic inducement. To further correct consumer attitudes to water conservation, energy conservation, recycling and pollution control the project should be extended through a follow-up test of the 'take-home' effect of these measures.

Considering that planning and implementation of major upgrading projects for metropolitan utilities, such as for water resource, takes at least a decade, a 'payback' approach to planning would assist in the decisions regarding policy. A longer-term vision should become mandatory for all projects in urban situations, where demand for services is outstripping existing infrastructure and requiring further large capital investment.

REFERENCES

1. A. C. Twort, D. D. Ratnayaka, Malcolm J. Brandt. (2000). *Water Supply.* London: Butterworth-Heinemann.
2. American Water Works Association. (2003). *Water Transmission and Distribution.* New York: American Water Works Association.
3. Armando Di Nardo, Michele Di Natale, Anna Di Mauro. (2013). *Water Supply Network District Metering: Theory and Case Study.* Berlin: Springer Science & Business Media.
4. Ashok K. Sharma, Donald Begbie, Ted Gardner. (2015). *Rainwater Tank Systems for Urban Water Supply.* London: IWA Publishing.
5. Bambos Charalambous, Chrysi Laspidou. (2017). *Dealing with the Complex Interrelation of Intermittent Supply and Water Losses.* London: IWA Publishing.
6. Bank, W. (2010). *World Development Indicators 2010.* New York: World Bank Publications.
7. Butler, D. (2003). *Advances in Water Supply Management: Proceedings of the International Conference on Computing and Control for the Water Industry, 15-17 September 2003, London, UK.*

London: Taylor & Francis.

8. Cabrera, E. (2003). *Pumps, Electromechanical Devices and Systems Applied to Urban Water Management, Volume 1.* London: Taylor & Francis.
9. Carlos A Iglesias, Luis Garrote, Antonio Cancelliere, Francisco Cubillo, Donald A. Wilhite. (2009). *Coping with Drought Risk in Agriculture and Water Supply Systems: Drought Management and Policy Development in the Mediterranean.* Berlin: Springer Science & Business Media.
10. Casey Brown, M. N. (2013). *Managing Climate Risk in Water Supply Systems.* London: IWA Publishing.
11. Cecilia Tortajada, Asit K. Biswas, Avinash Tyagi. (2014). *Water Management and Climate Change: Dealing with Uncertainties.* London: Routledge.
12. David Butler, Fayyaz Ali Memon. (2005). *Water Demand Management.* London: IWA Publishing.
13. Development, I. I. (1999). *PLA Notes 35: Community Water Management.* London: IIED.
14. DIANE. (1978). *Guidance manual for conducting sanitary surveys of public water systems surface water and ground water under the direct influence (GWUDI).* London: DIANE Publishing.
15. El-Naser, H. (2009). *Management of Scarce Water Resources: A Middle Eastern Experience.* New York: WIT Press.
16. Emily Alcott, Mark S Ashton, Bradford S Gentry. (2013). *Natural and Engineered Solutions for Drinking Water Supplies: Lessons from the Northeastern United States and Directions for Global Watershed Management.* New York: CRC Press.
17. Enrique Cabrera, Jorge García-Serra. (2013). *Drought Management Planning in Water Supply Systems: Proceedings from the UIMP International Course held in Valencia, December 1997.* Berlin: Springer Science & Business Media.
18. Enrique Cabrera, R. Cobacho, Jay Lund. (2002). *Regional Water System Management: Water Conservation, Water Supply and System Integration.* New York: CRC Press.

19. Grigg, N. S. (2010). *Governance and Management for Sustainable Water Systems.* London: IWA Publishing.
20. Gurewitz, P. H. (1978). *Hydraulic research in the United States and Canada, 1976.* New York: U.S. Dept. of Commerce, National Bureau of Standards.
21. Helena Alegre, Jaime M. Baptista, Enrique Cabrera Jr, Francisco Cubillo, Patricia Duarte, Wolfram Hirner, Wolf Merkel, Renato Parena. (2016). *Performance Indicators for Water Supply Services.* London: IWA Publishing.
22. International Association of Hydrological Sciences. Scientific Assembly. (2005). *Sustainable Water Management Solutions for Large Cities.* London: INTERNATIONAL ASSOCIATION OF HYDROLOGICAL SCIENCES.
23. Jaroslav Pollert, Bozidar Dedus, NATO Public Diplomacy Division. (2006). *Security of Water Supply Systems: From Source to Tap.* Berlin: Springer Science & Business Media.
24. Keith W. Hipel, Liping Fang, Johannes Cullmann, Michele Bristow. (2015). *Conflict Resolution in Water Resources and Environmental Management.* Berlin: Springer.
25. Kiyoshi Kobayashi, Ismu Rini Dwi Ari, Isabel C. Escobar, Andrea Schaefer. (2014). *Community Based Water Management and Social Capital.* London: IWA Publishing.
26. Lindsay Knight, World Health Organization. (2003). *The Right to Water.* Switzerland: World Health Organization.
27. Luis Santos Pereira, Ian Cordery, Iacovos Iacovides. (2009). *Coping with Water Scarcity: Addressing the Challenges.* Berlin: Springer Science & Business Media.
28. Mehta, A. (2011). *Toolkit for Public–Private Partnerships in Urban Water Supply for the State of Maharashtra, India.* Phillipines: Asian Development Bank.
29. National Research Council, Division on Earth and Life Studies, Commission on Geosciences, Environment and Resources, Committee on Small Water Supply Systems. (1997). *Safe Water From Every Tap: Improving Water Service to Small Communities.* New York: National Academies Press.

30. National Research Council, Division on Earth and Life Studies, Commission on Geosciences, Environment and Resources, Committee to Review the New York City Watershed Management Strategy. (2000). *Watershed Management for Potable Water Supply: Assessing the New York City Strategy.* New York: National Academies Press.
31. Organization, W. H. (2002). *Guidelines for Drinking-water Quality: Microbiological Agents in Drinking Water. Addendum.* Switzerland: World Health Organization.
32. Organization, World Health. (2004). *Guidelines for Drinking-water Quality: Recommendations.* Switzerland: World Health Organization.
33. Peter Harvey, Bob Reed. (2004). *Rural Water Supply in Africa: Building Blocks for Handpump Sustainability.* London: WEDC, Loughborough University.
34. Peter Lonsdale, Dusan Obradovic. (1998). *Public Water Supply: Models, Data and Operational Management.* New York: CRC Press.
35. Petr Hlavinek, Cvetanka Popovska, Ivana Mahrikova, Tamara Kukharchyk. (2009). *Risk Management of Water Supply and Sanitation Systems.* Berlin: Springer Science & Business Media.
36. Ratnayaka, Don D. Ratnayaka, Malcolm J. Brandt, Michael Johnson. (2009). *Water Supply.* London: Butterworth-Heinemann.
37. Sharad K. Jain, V. S. (2003). *Water Resources Systems Planning and Management.* Netherlands: Elsevier.
38. Simonovic, S. P. (2012). *Managing Water Resources: Methods and Tools for a Systems Approach.* London: Routledge.
39. Sobsey, M. D., Water, S., & World Health Organization. (2002). Managing water in the home. *accelerated health gains from improved water supply.*
40. Stephenson, D. (2012). *Water Supply Management.* Berlin: Springer Science & Business Media.
41. Stuart Hamilton, Ronnie McKenzie. (2014). *Water Management and Water Loss.* London: IWA Publishing.
42. Takizawa, S. (2008). *Groundwater Management in Asian Cities: Technology and Policy for Sustainability.* Berlin: Springer Science

& Business Media.

43. The World Bank. (2010). *Water and Development: An Evaluation of World Bank Support, 1997-2007.* New York: World Bank Publications.
44. UNICEF. (2006). *Progress for Children: A Report Card on Water and Sanitation.* New York: UNICEF.
45. UNICEF Staff. (2010). *UNICEF Annual Report 2009.* New York : UNICEF .
46. United Nations Environment Programme. (2000). *UNEP Water Policy and Strategy.* New York: UNEP/Earthprint.
47. United States. National Bureau of Standards. (1976). *Hydraulic research in the United States and Canada, Issue 497.* New York: U.S. Govt. Print.
48. Vijay P. Singh, Ram Narayan Yadava. (2003). *Watershed Management.* New Delhi: Allied Publishers.
49. WHO/UNICEF Joint Water Supply and Sanitation Monitoring Programme. (2014). *Progress on Drinking Water and Sanitation: 2014 Update.* Switzerland: World Health Organization.
50. William Whipple, Jr. (1994). *New Perspectives in Water Supply.* New York: CRC Press.
51. World Bank . (2010). *Results and Performance of the World Bank Group: IEG Annual Report 2010, Volume 1.* New York: World Bank Publications.
52. World Bank. (2009). *World Development Report 2010: Development and Climate Change.* New York: World Bank Publications.
53. World Health Organization. (2009). *Water Safety Plan Manual: Step-by-step Risk Management for Drinking-water Suppliers.* Switzerland: World Health Organization.
54. World Health Organization. (2010). *Training Workbook on Water Safety Plans for Urban Systems.* Switzerland: World Health Organization.
55. World Health Organization. (2010). *World Health Statistics 2010.* Switzerland: World Health Organization.
56. World Health Organization. (2013). *Progress on Sanitation and Drinking-water.* Switzerland: World Health Organization.

57. World Health Organization. (2015). *Progress on Sanitation and Drinking Water : 2015 Update and MDG Assessment.* Switzerland: World Health Organization.
58. World Water Assessment Programme (United Nations). (2009). *Water in a Changing World, Volume 1.* London: Earthscan.
59. World Water Assessment Programme. (2012). *The United Nations World Water Development Report 3: Water in a Changing World (Two Vols.).* London: Routledge.

INDEX

W